ECO-DRIVING

FROM STRATEGIES TO INTERFACES

Transportation Human Factors:
Aerospace, Aviation, Maritime, Rail, and Road Series

Series Editor
Professor Neville A Stanton
University of Southampton, UK

Automobile Automation: Distributed Cognition on the Road
Victoria A. Banks, Neville A. Stanton

Eco-Driving: From Strategies to Interfaces
Rich C. McIlroy, Neville A. Stanton

ECO-DRIVING

FROM STRATEGIES TO INTERFACES

Rich C. McIlroy and Neville A. Stanton

CRC Press
Taylor & Francis Group
Boca Raton London New York

CRC Press is an imprint of the
Taylor & Francis Group, an **informa** business

CRC Press
Taylor & Francis Group
6000 Broken Sound Parkway NW, Suite 300
Boca Raton, FL 33487-2742

International Standard Book Number-13: 978-1-138-03201-9 (Hardback)

Library of Congress Cataloging-in-Publication Data

Names: McIlroy, Rich C., author. | Stanton, Neville A. (Neville Anthony), 1960- author.
Title: Eco-driving : from strategies to interfaces / Rich C. McIlroy, Neville A. Stanton.
Description: Boca Raton : Taylor & Francis, CRC Press, 2018. | Series: Transportation human factors : aerospace, aviation, maritime, rail, and road series | Includes bibliographical references.
Identifiers: LCCN 2017019281 | ISBN 9781138032019 (hardback : alk. paper)
Subjects: LCSH: Automobile driving--Energy conservation. | Automobile driving--Environmental aspects. | Automobile driving--Human factors.
Classification: LCC TL151.5 .M35 2017 | DDC 629.28/30286--dc23
LC record available at https://lccn.loc.gov/2017019281

Visit the Taylor & Francis Web site at
http://www.taylorandfrancis.com

and the CRC Press Web site at
http://www.crcpress.com

Dedication

For Ire – Rich

For Maggie, Josh, and Jem – Neville

Contents

Preface

The work presented in this book, representing the output of five years of research, was initially motivated by two broad factors: a belief in the society-wide need to reduce global resource consumption, and an interest in ecological interface design, and the skills, rules and knowledge taxonomy, to both describe human behaviour and to inform design. At the beginning of the research journey, the low-carbon vehicle domain (e.g. hybrid and electric vehicles) was chosen as an appropriate field within which to address these two factors. As the reader will discover, this soon developed into a broader eco-driving focus (i.e. the driving behaviours associated with low energy consumption), across not only low-carbon vehicles but also road vehicles of any type. One aim of this book, therefore, is to shed further light on this increasingly studied topic. This is done so via the use of a variety of human factors methods; as will be seen, some of these methods were more enlightening than others.

With regard to the theoretical motivation to the research, rather than attempt to apply the full ecological interface design method to in-vehicle interface design, the focus was narrowed to include only a small subset of its core principles. This led to the exploration of the theoretical justifications for the use of in-vehicle haptic stimuli for the support of certain in-vehicle behaviours. This book therefore provides the interested reader with a relatively in-depth discussion of the ability of multisensory information to support behaviour at different levels of cognitive control, a key concern of the ecological interface design methodology.

Finally, this book has an additional, purely practical aim, to provide those working in the automotive vehicle industry with advice on how to help drivers make the most out of their vehicle's energy reserves, in whatever form that energy may be stored. To this end, we provide an argument for the support of anticipatory behaviours in the vehicle, the uptake of which would not only be beneficial for fuel economy (the focus of this book), but also for safety. The technologies required to make such a system as that described in the latter chapters of this book are not far away, and it is now, in the earliest stages of product development, that the application of human factors can provide most benefit.

Acknowledgements

We would first like to acknowledge the Engineering and Physical Sciences Research Council (EPSRC) and Jaguar Land Rover. Without their financial support this research would not have been possible.

We must also thank Antony Wood and Louise Godwin who, at the time, were based at the Institute of Sound and Vibration Research's electronics workshop. Without their help and expertise we would have had no in-vehicle system to test.

Paul Salmon and Miles Thomas at the University of the Sunshine Coast, Queensland, also deserve considerable thanks for providing us with a significant amount of data. Although we did not enjoy the arduous task of transcribing it, without said data this book would be one chapter shorter.

And finally, to all the people who participated in the simulator studies, responded to the survey and allowed us to interview them. We did not pay any of you for your time, yet you still provided it – for this, we thank you. Without such people none of this research (or, indeed, the vast majority of human-based research) would have been possible at all. And for those in whom we induced unbearable sickness in the simulator and had to leave early, we are sorry, and thank you for not soiling the car.

Authors

Dr Rich C. McIlroy received a BSc (Hons) degree in psychology and an MSc degree in research methods in psychology from the University of Southampton, UK, in 2008 and 2009, respectively. He was recently awarded his engineering doctorate by the same university, having been based in the Transportation Research Group, Faculty of Engineering and the Environment. He has published over 16 articles across a variety of topics including eco-driving and the effect of in-vehicle information on driving behaviour and fuel use, the effect of multisensory information on responding, the general utility of ecological interface design, the link between expertise development and verbal reporting, the use of non-intrusive verbal reporting for information acquisition, and the ability of cognitive work analysis and its various components to support decision-making and system design in a variety of domains, from rail transport to system requirements specification.

Professor Neville A. Stanton, PhD, DSc, is a chartered psychologist, chartered ergonomist and chartered engineer. He holds the Chair in Human Factors Engineering in the Faculty of Engineering and the Environment at the University of Southampton in the United Kingdom. He has degrees in occupational psychology, applied psychology and human factors engineering and has worked at the Universities of Aston, Brunel, Cornell and MIT. His research interests include modelling, predicting, analysing and evaluating human performance in systems as well as designing the interfaces and interaction between humans and technology. Professor Stanton has worked on the design of automobiles, aircraft, ships and control rooms over the past 30 years, on a variety of automation projects. He has published 35 books and over 280 journal papers on ergonomics and human factors. In 1998, he was presented with the Institution of Electrical Engineers Divisional Premium Award for research in system safety. The Institute of Ergonomics and Human Factors in the United Kingdom awarded him the Otto Edholm Medal in 2001, the President's Medal in 2008 and the Sir Frederic Bartlett Medal in 2012, which were awarded for his contributions to basic and applied ergonomics research. The Royal Aeronautical Society awarded him and his colleagues the Hodgson Prize in 2006 for research on design-induced, flight-deck error published in *The Aeronautical Journal*. The University of Southampton awarded him a Doctor of Science in 2014 for his sustained contribution to the development and validation of human factors methods.

Abbreviations

AA	Automobile Association
ACC	Automatic Cruise Control
ADAS	Advanced Driver Assistance System
ADS	Abstraction Decomposition Space
AH	Abstraction Hierarchy
ANOVA	Analysis of Variance
BA	Bachelor of Arts
BBC	British Broadcasting Corporation
BEV	Battery Electric Vehicle
BSc	Bachelor of Science
BWR	Boiling Water Reactor
CAN	Controller Area Network
CO₂	Carbon Dioxide
ConTA	Control Task Analysis
CWA	Cognitive Work Analysis
DECC	Department of Energy and Climate Change
DfT	Department for Transport
DL	Decision Ladder
DMI	Direct Manipulation Interfaces
DURESS	Dual Reservoir System Simulation
EID	Ecological Interface Design
EU	European Union
FIT	Feedback Intervention Theory
GCSE	General Certificate of Secondary Education
GPS	Global Positioning System
HEV	Hybrid Electric Vehicle
HMI	Human Machine Interface
HTA	Hierarchical Task Analysis
Hz	Hertz
ICE	Internal Combustion Engine
ICU	Intensive Care Unit
IPCC	Intergovernmental Panel on Climate Change
KBB	Knowledge-Based Behaviour
KM	Kilometres
KMH	Kilometres per Hour
MANOVA	Multivariate Analysis of Variance
MPG	Miles per Gallon
MPH	Miles per Hour
ORCa	On Road Capability
PCM	Perceptual Cycle Model
PHEV	Plug-in Hybrid Electric Vehicle
RBB	Rule-Based Behaviour

REEV	Range-Extended Electric Vehicle
RPM	Revolutions per Minute
SBB	Skill-Based Behaviour
SOCA	Social Organisation and Cooperation Analysis
SRK	Skills, Rules and Knowledge
StrA	Strategies Analysis
UAV	Unmanned Aerial Vehicle
UK	United Kingdom (of Great Britain and Northern Ireland)
US	United States (of America)
USAF	United States Air Force
VPA	Verbal Protocol Analysis
WCA	Worker Competencies Analysis
WDA	Work Domain Analysis

1 Introduction

1.1 BACKGROUND

The research presented in this book was motivated, in the main part, by two principal factors: (1) a belief in the necessity to protect the environment we inhabit through the minimisation of our usage of the planet's natural resources and (2) an interest in the ability of a particular theoretical taxonomy to both describe human behaviour and cognition, and to inform system design. The combination of these two motivational forces (alongside a number of other less significant, yet nonetheless important influences) guided the overarching focus of the research presented in the coming pages: the encouragement and support of eco-driving in the private road vehicle.

The first point above stems from the growing concern surrounding anthropometrically caused climatic change (IPCC 2014), and the issue of sustainability (World Commission on Environment and Development 1987). As shall be discussed in more detail in subsequent chapters of this book, it is the transport domain in particular that is lagging behind, with other sectors (e.g. domestic, industry) showing far greater improvements in energy use and emissions reductions (Department of Energy and Climate Change 2012a). Indeed, despite a 24% decrease in *total* emissions between 1990 and 2009 across the EU, transport's emissions rose by 29% (Hill et al. 2012).

Moreover, when looking at transport's share of resource consumption and emission volumes more closely, we find that it is private road transport that features most highly. Across the EU in 2012, road transport accounted for 17.5% of *all* greenhouse gas emissions, emissions that include those from all forms of transport, industry, domestic use, agriculture, and electricity production (European Commission 2015). Although we have seen a decrease in emission volumes over the past 7 years, levels are still 20.5% above those seen in 1990 (ibid.).

Private road transport, that is, the cars in which we travel to and from work, to visit relatives, or to take the kids to school (e.g.), plays an especially significant role, accounting for more than half of all the emissions from transport in the UK (Commission for Integrated Transport 2007). There has, in the past 5 years or so, been a significant increase in the number of hybrid and electric vehicles registered in the UK (Society of Motor Manufacturers and Traders 2016) (Figure 1.1). In Europe at least (The Shift Project 2015), these types of vehicles certainly contribute to reductions in energy usage and emission volumes across their lifespan (Hawkins et al. 2013); however, alongside opportunities, these vehicles, by nature of both their novelty and their complexity (particularly for hybrids, in which two different fuel systems and/or drivetrain technologies are present), give rise to a number of challenges (see Chapter 2).

There is no doubt that technological advancement, in both vehicles and infrastructure, has a huge part to play in our journey towards a fully sustainable transport system.

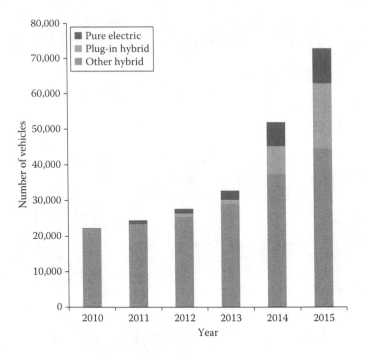

FIGURE 1.1   Total number of pure electric, plug-in hybrid, and other hybrid vehicles (i.e. non-plug-in) registered in the UK annually. (Data from http://www.smmt. co.uk/category/news-registration-evs-afvs)

This is not, however, the only way in which sustainability can be achieved, and it is not the focus of this book. Rather, the work presented herein approaches the problem from a behavioural perspective. The question is: how can we help people make the behavioural changes necessary to take full advantage of these new lower-emission technologies?

The reader will see in Chapter 2 that the initial focus of the research project described in this book was on low-carbon technologies, namely hybrid and electric vehicles. However, as will be discussed, simply buying a hybrid or electric car does not automatically make for a sustainable transport system; the way in which it is driven is also important. Of course, not driving at all is arguably the most sustainable way to reduce emissions; however, it is a flight of fancy to expect all drivers to suddenly abandon their cars in favour of walking or cycling for all of their journeys. A more realistic goal would be to aim for the widespread adoption of sustainable behaviours *in* the vehicle. When we consider that the average driver could save around 10% of the fuel they use simply by modifying the way in which they drive (Barkenbus 2010), the significance of the total potential energy and emission savings that would result if every driver were to adopt a fuel-efficient driving style becomes abundantly clear.

Although one could argue that the adoption of an economical driving style is especially important in electric vehicles (given, e.g. the need to deal with the range limitations not inherent to vehicles equipped with an internal combustion engine

(ICE) (see Chapter 2)), such a practice can result in fuel savings in any road vehicle. From Chapter 3 onwards, this book therefore focuses not on hybrid and electric vehicles, but on the behaviours that characterise fuel-efficient driving in *any* private road vehicle. These behaviours are collectively known as 'eco-driving', and are central to this body of work. The primary question addressed in this book is how to best encourage the uptake of such behaviours. In other words, how do we help drivers behave in a more fuel-efficient manner when in control of the vehicle? As will be discussed, there are a variety of ways in which this can be done, from pre-trip eco-driving training to post-trip presentation of energy consumption figures (Barkenbus 2010). This book is focused on just one; the provision of in-vehicle information, presented concurrently with the driving task.

In the following chapter more time is devoted to discussing the importance of in-vehicle information design. As aforementioned, the chapter pays particular attention to low-emission vehicles, and the potential for the encouragement of new, fuel-efficient driving habits. This is not simply a question of efficiency, but also safety. When adding information to the in-vehicle environment, care must be taken to ensure that it does not negatively affect performance of the primary driving task e.g. through increasing workload or causing distraction (e.g. Harvey et al. 2011a). The design of the information, therefore, is critical.

In the latter part of Chapter 2, ecological interface design (Rasmussen and Vicente 1989; Vicente and Rasmussen 1992) is introduced, and argued to be a potentially promising method for the design of an in-vehicle information system. We will not go into great detail here in describing the method; significant time is devoted to the topic in later chapters. For the purposes of this introductory chapter, however, it is useful to outline its core principles, and how these have shaped this research project.

Ecological interface design is partly based on the skills, rules and knowledge taxonomy of human behaviour (Rasmussen 1983), the theoretical taxonomy alluded to in this chapter's opening paragraph. The three terms describe the levels of cognitive control with which an actor interacts with the environment; skill-based behaviour involves automatic, direct interaction; rule-based behaviour involves associating familiar perceptual cues in the environment with stored rules for action and intent; knowledge-based behaviour involves analytical problem-solving based on symbolic reasoning and stored mental models. The ecological interface design method aims to produce an interface that supports behaviour at all three levels of cognitive control, by supporting interaction via time-space signals (for skill-based behaviour), by providing consistent mapping between constraints in the environment and cues at the interface (for rule-based behaviour), and by representing the system's structure via an externalised mental model (for knowledge-based behaviour).

In the early stages of this project, ecological interface design was considered as an appropriate methodology for the design of in-vehicle systems for alternative drivetrain vehicles due to its ability to help in the design of systems that support the development of accurate mental models of complex systems, allowing for behaviour at all three levels of cognitive control. As the research progressed, however, the focus shifted away from low-carbon vehicles specifically, and also began to concentrate on the first of the design method's three principles; to support interaction via time-space signals in order to encourage behaviour at the skill-based level. This shift was inspired, in part, by research reported by Birrell, Young and colleagues (Birrell, Fowkes, et al. 2014; Birrell and Young 2011; Young et al. 2011). These articles reported on an in-vehicle interface, designed using the principles of ecological interface design, which not only attempted to display domain constraints, but also to provide information on the specific ways in which drivers could alter their behaviour to improve safety and fuel economy.

This concept, of guiding the fuel-efficient behaviours themselves (as opposed to attempting to provide an externalised model of the system), gave direction to the information gathering activities and experimental work presented in Chapters 6 to 9 of this book. As shall be discussed in the coming chapters, the *expert* eco-driver performs the task in a way that approaches automaticity, that is, they are performing at the skill-based level of cognitive control. One of the questions that guided the design of the information system described in Chapter 7, and the design of the experiment described in Chapter 8, was whether or not vibrotactile information, presented at the site of control (see Chapters 5 and 6), can support this type of responding in the *novice* eco-driver, that is, can it support eco-driving behaviours at the skill-based level of cognitive control? Not only did this present some interesting practical questions (regarding the actual fuel saved with use of such a system, and the acceptance of that system by participants), but also presented a number of theoretical issues regarding the ways in which multisensory individuals interact with their multi-modal environment.

1.2 AIMS AND OBJECTIVES

The main aim of this research project was to investigate the potential for in-vehicle information to support eco-driving in the road vehicle (be that a fully electric vehicle, a hybrid, or one equipped only with an ICE) in a way that neither increases workload nor distracts the driver from the primary driving task and, additionally, in a way that drivers are willing to accept and use. It is important to state that this research is not an investigation of the psychophysiological effects of stimuli of differing intensities and frequencies, nor is it a technically focused description of the algorithms and computations required to integrate information from radar, GPS or a vehicle's CAN bus in order to provide stimuli for the encouragement of eco-driving. For more information regarding the first of these research areas the reader is directed to work by, for example, Michael Griffin and colleagues of the University of Southampton's Institute of Sound and Vibration Research (Forta et al. 2011; Gu and Griffin 2012). For the integration of topographical and digital map data with sensor and engine data, the reader is referred to work surrounding Continental's eHorizon project (Continental 2015). This project involves the optimisation of engine control, transmission control and, importantly, driver assistance systems, via the use of information regarding the stretch of road ahead of the vehicle (Varnhagen and Korthaus 2010).

The research presented in this book is an investigation of the effects on human behaviour, and on system acceptance, of the kinds of in-vehicle information that are either currently available, or that are likely to be possible in the near future and, additionally, of how best to present that information. Furthermore, there was an aim to investigate the potential for in-vehicle information provided at the site of control (i.e. through the accelerator pedal, as will be revealed in the latter half of this book) to support skill-based behaviour in the novice eco-driver. This first aspect, simply to encourage fuel-efficient use of the vehicle, provides the more practically focused side of this book; the second aspect, related to the skills, rules and knowledge taxonomy, presents the theoretical aspect.

1.3 BOOK OUTLINE

This book is organised into 10 chapters, this introductory chapter being the first. Below, each of the remaining nine chapters is introduced in turn.

Chapter 2: Design, Behaviour and Energy Use

This provides the backdrop to the book by bringing together various strands of research, including the effect of the design of a technological object on behaviour, the inter-related nature of goals and feedback in guiding performance, the effect on fuel economy of different driving styles, and the various challenges brought about by hybrid and electric vehicles including range anxiety, workload and distraction, complexity and novelty. This chapter also introduces ecological interface design, arguing it to be well suited to deal with the novelty of the low-carbon vehicle, particularly through its ability to support the development of accurate mental models of the system. The discussion is couched in terms of the support of energy-efficient use of the vehicle.

Chapter 3: Driving and the Environment: An Exploratory Survey Study

This chapter is concerned with the general public's knowledge and perceptions of eco-driving as a practice, their awareness of and propensity to perform specific eco-driving behaviours, and the relationships these variables have with demographics (both general and driving-specific) and environmental attitudes. A survey of 321 respondents revealed that the majority are aware of eco-driving and have a positive attitude towards it; however, knowledge of the specific behavioural strategies for fuel-efficient driving was not high. Although relationships were found between energy use attitudes and both knowledge of and propensity to perform eco-driving behaviours, these relationships were weak.

Chapter 4: Verbal Reports: An Exploratory On-road Study

In order to begin to understand the actual behaviours exhibited, and cognitive structures held by *individual* drivers, Ericsson and Simon's verbal protocol analysis technique (Ericsson and Simon 1980, 1993) was applied in an on-road setting. Twenty participants each drove a 15 to 20 minute route, during which they were required to 'think aloud'. The transcripts of 19 of these participants were transcribed verbatim and a coding scheme iteratively developed, partly based on theory (i.e. top-down), partly on the content of the transcripts themselves (i.e. bottom-up). The coding scheme was then applied to all the transcripts, thereby categorising each identifiable unit of speech into the various codes. Objective vehicle data were also recorded, at 10 Hz, and included measures such as vehicle speed and accelerator pedal position. Although every effort was made to link objectively measured driving behaviours with the content of the transcripts, no relationships could be found.

Chapter 5: Two Decades of Ecological Interface Design, and the Importance of the SRK Taxonomy

In a momentary departure from the driving focus of the book, this chapter deals only with ecological interface design, providing a review of the past two decades of the method's applications published in the academic literature. The method is described in more detail, and the importance of the skills, rules and knowledge (SRK) taxonomy to the framework is specifically discussed following the finding that 40% of reviewed applications do not cite this component, despite its centrality to the method.

Chapter 6: A Decision Ladder Analysis of Eco-driving: The First Step Towards Fuel-efficient Driving Behaviour

This chapter draws heavily on the SRK taxonomy in a decision ladder analysis of eco-driving, discussing results in terms of how this can inform the design of an in-vehicle, eco-driving support system. A review was conducted of the academic literature, and of more publicly available resources (i.e. free to access, those not requiring subscription), identifying four largely distinct driving activities that each play a central role in the use of fuel in the private road vehicle. A focus group involving four researchers in the transport ergonomics field, followed by a series of five interviews with eco-driving experts, served to validate, supplement and further specify the models.

Chapter 7: In-vehicle Information System Design

Based on the arguments arising from the decision ladder analysis of eco-driving presented in Chapter 6, a system was developed that aimed to encourage fuel-efficient driving in the novice eco-driver; this chapter describes the design process of that system, and the resulting components and functions. The chapter also provides information regarding the Southampton University driving simulator, and presents results of the pilot testing of the system and of the driving scenarios that were to be used in the experiment described in Chapter 8.

Chapter 8: Ecological Driving with Multisensory Information

This chapter presents the first experimental evaluation of the in-vehicle eco-driving support system described in Chapter 7. Behaviour when driving 'normally' was compared to that exhibited when participants were asked to drive economically, and to that exhibited when provided with feedback in three sensory modes (audition, vision, touch), individually and in all combinations thereof. Results suggested that participants were already largely aware that harsh acceleration is to be avoided when eco-driving; however, significantly greater coasting distances (when approaching slowing events) were seen only under conditions of feedback. Few differences were seen between the different sensory modes and combinations; however, for some measures visual-only information was shown to be less effective than combinations involving auditory and vibrotactile stimuli. Although it encouraged compliance, the auditory stimulus was not well received by participants. Results are discussed in terms of the ability of feedback in different sensory modes to support eco-driving in different drivers, and in relation to the SRK taxonomy.

Chapter 9: When to Give Those Good Vibrations

In the second experimental analysis of the eco-driving support system, only haptic (vibrotactile) information was investigated. The research presented in this chapter had a more practical focus (rather than theoretical), and investigated only the encouragement of coasting when approaching slowing or stopping events. The simulator study assessed the effects of three different time-to-event stimulus timings on objective driving performance, and on subjective measures of acceptance, ease of use and intention to use. The shortest time-to-event had a marginally damaging effect on performance, and was not well received by participants. Both medium and long time-to-event stimuli performed well on subjective measures, and both facilitated increased eco-driving performance. The longest lead-time stimulus was the most effective, resulting in 11% fuel savings compared to baseline. Findings are discussed in terms of the importance of the timing of information, and regarding the need for longer-term research on the potential effects of system failure on performance and safety.

Chapter 10: Conclusions

The final chapter of this book summarises the work undertaken and described in the preceding eight chapters. Methodological, practical and theoretical contributions are outlined, implications of the research are discussed, and avenues for future work are suggested.

1.4 CONTRIBUTION TO KNOWLEDGE

The work presented in the coming chapters contributes, to varying degrees, to our understanding of eco-driving as a means for reducing the impact of road transport on the environment, to the literature concerning haptic information in the vehicle, and to the theory underlying the first of ecological interface design's three principles; to support skill-based behaviour with time-space signals. Regarding the first point, it was already clear from the existing literature that eco-driving can have a significant, beneficial effect on energy use in the vehicle. The research described in this book adds to extant knowledge by demonstrating the effectiveness of a method by which these benefits might be realised; namely, to provide information that directly supports smoother acceleration profiles and increased coasting behaviours (two behaviours identified in this book to be of particular significance in eco-driving). This is in contrast to the majority of previous research that provides feedback regarding *current* energy usage levels, or information detailing the vehicle's remaining energy reserves. Results of the experimental work (Chapters 8 and 9) led to the further suggestion that focusing solely on the support of coasting may be more suitable (in terms of acceptance and effectiveness) than attempting to support both enhanced coasting behaviours *and* smooth acceleration.

With regard to the second point above, this book adds to the body of knowledge surrounding the effects of accelerator-based haptic feedback in the vehicle by comparing the effects of information presented across different sensory modes. Though such comparisons have, in the past, been made between haptic and visual information, this book goes further by also looking at auditory information, suggesting that vibrotactile information is as effective as auditory (in encouraging compliance; visual being less effective), but with far higher user acceptance. This book also investigates a vibrotactile haptic stimulus rather than force- or stiffness-feedback, methods far more commonly reported in the literature. Additionally, the timing of the coasting advice, that is, the distance ahead of a slowing event at which information suggesting removal of the foot from the accelerator pedal is presented, is shown to be important for both system effectiveness (in reducing fuel consumption) and for user acceptance.

Finally, in terms of the contributions to the ecological interface design and SRK theory, this book provides a thorough review of the past two decades of the design method's applications, argues for the importance of the SRK framework as a fundamental part of the method, and significantly furthers the discussion of the ability of haptic information, provided at the site of control, to support behaviour at different levels of cognitive control. This final point is of particular significance when we consider that the vast majority of research surrounding ecological interface design, and indeed the SRK taxonomy, be it theoretical or applied, focuses almost exclusively on visual interfaces (with a small number of notable exceptions, as will be discussed). Although results from the experiments described in the latter part of this book cannot definitively answer all of the questions arising from the discussions presented herein, headway has been made.

2 Design, Behaviour and Energy Use

2.1 INTRODUCTION

Despite a small number of sceptics (Reser et al. 2011), it is now largely accepted that anthropometric sources, that is, humans past and present, are the primary cause of the earth's rising temperature (Intergovernmental Panel on Climate Change 2007). We, as a 7 billion-strong collection of energy-using individuals, are constantly consuming more and more energy and resources to satisfy our daily needs, and the planet cannot indefinitely support our current level of resource usage let alone projected future consumption rates should prevailing trends continue (International Energy Authority 2012).

With this in mind, the aim of this chapter is to bring to attention an important avenue for the mitigation of climate change and the reduction in both the usage of resources and the emission of environmentally damaging by-products; namely, the design of technological objects, specifically battery-only, and hybrid–electric private road vehicles. The review is intended to highlight the importance of the manner in which these technological objects are used, and how ergonomics can be applied not only to support safety and enhance usability, but also to encourage reductions in energy consumption (and, in turn, waste production).

Transport's role in the global warming issue will be examined, followed by a discussion on the influence of design on behaviour, both generally and, more specifically, in terms of vehicle usage. The usability and safety of in-vehicle systems will be discussed, followed by a brief examination of a particular analysis and design framework that can offer the basis from which to design a driver–vehicle interface that will ensure usability and encourage energy conservation behaviours while not detracting from the goal of ensuring safety. First, it is important to provide some background information regarding our overuse of the planet's resources.

2.2 SUSTAINABILITY AND TRANSPORT

The issue of sustainability does not only concern our environment's ability to provide resources, but also its ability to absorb waste (World Commission on Environment and Development 1987). It is primarily the emission of the waste product carbon dioxide (CO_2 – the by-product of using fossil fuels as a primary energy source), emitted in volumes that our environmental system does not have the capacity to absorb, that is causing the observed increases in our environment's temperature (Intergovernmental Panel on Climate Change 2007). As of 2011, petroleum accounted for 48% of total final energy consumption in the UK (Figure 2.1; Department of Energy and Climate Change 2012a). Though progress has been made in other sectors (e.g. industry, domestic, commercial), transport is lagging behind in sustainability terms.

For example, though CO_2 emissions from non-transport sources fell by almost 23% between 1990 and 2010, emissions from the UK transport sector increased marginally (Department of Energy and Climate Change 2012b). The issue is especially relevant for private transport given that, in the UK, 54% of all transport's carbon emissions (including those from air, rail, shipping and all private and commercial road transport) were produced by cars (Figure 2.2; Commission for Integrated Transport 2007).

The importance of road transport cannot be underestimated; it 'underpins our way of life' (King 2007, 3), supporting the high level of personal mobility to which the vast majority of us have become accustomed. Not only do we rely on the road transport system to get us around, we specifically design our built environment based on the constraints of road vehicles. Furthermore, private road transport still offers the only form of motorised travel that transports us from door to door, is entirely flexible regarding departure time and destination, and is often the fastest mode for distances up to 500 km (Damiani et al. 2009).

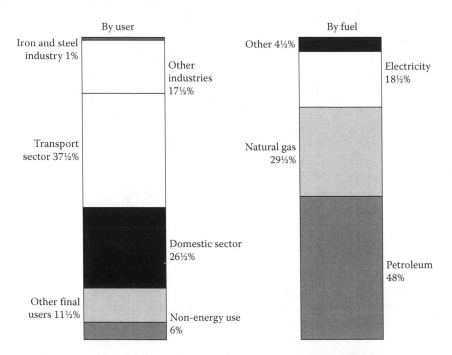

FIGURE 2.1 Breakdown of final energy consumption in the UK in 2011. Note that 'fuel' refers to the final way in which energy is consumed by the user, hence the inclusion of electricity. (Adapted from Department of Energy and Climate Change, *Digest of United Kingdom energy statistics*, The Stationary Office, London, 2012a.)

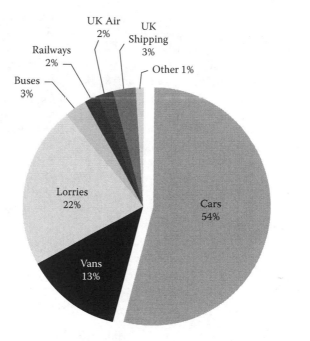

FIGURE 2.2 Transport emissions by mode in the UK. (Adapted from Commission for Integrated Transport, *Transport and Climate Change*, The Stationary Office, London, 2007.)

Though we may be able to encourage people to make fewer journeys (e.g. by encouraging working from home), and to improve public transport services through investment, our reliance on private road transport, and indeed the fundamental requirement for travel as a whole, makes it unrealistic to assume that this will be sufficient. Indeed, as Stanton et al. (2012) describe, the removal of the barriers to modal shift (i.e. getting drivers and passengers out of the private motor vehicle and on to public transport) is a highly complex, multifaceted issue that will not be easily remedied. It is therefore apparent that if we are to achieve the 80% reduction in CO_2 emissions posited by the UK Government (in their 2008 Climate Change Act) as necessary to avoid the most serious consequences of increasing the earth's temperature [both environmentally (Intergovernmental Panel on Climate Change 2007) and economically (Garnaut 2011; Stern 2006)] we will have to enact a wide variety of mitigation strategies. Hence the burgeoning interest in the electrification of private road transport.

2.3 SUSTAINABILITY AND ERGONOMICS

Technological advancement is of course a crucial part of reducing fossil fuel reliance; however, it is not the only challenge. We must also have behavioural change (Stern 2006). One avenue for the encouragement of this change is through the design of products. Consider this: consumers' behaviour is shaped by the product they are using, and the product they are using has been designed with a particular activity in mind (Stanton and Baber 1998). Design, therefore, shapes behaviour. In technical objects, the use phase of an item is often where the most significant environmental impact occurs (Lockton et al. 2008b); hence interaction design provides an avenue for energy or waste reduction (Lockton et al. 2008b). This is particularly significant in the transport domain, given that the vast majority of emissions and energy usage occurs at the point of use; life cycle analyses (i.e. those considering production, use and disposal) suggest that, for road vehicles, 76% of CO_2 emissions and 80%–90% of energy use can be attributed to the burning of fuel in an internal combustion engine (ICE) (Organisation of Economic Cooperation and Development 1993; see also Hawkins et al. 2013 for life cycle considerations of both traditional and low-carbon vehicles).

The connection between ergonomics and sustainability has been discussed by a number of academics within the domain, for example: Flemming et al. (2008) with their call for the application of ergonomics to sustainability; Martin et al. (2012) with their discussion on designing for sustainability; and Thatcher (2012) with his essay on 'green ergonomics' and the alignment of the goals of ergonomics with those of environmental sustainability. Although these discussions may not have been specifically targeted at transport, it has been recognised that the electric and hybrid vehicle domain offers a promising avenue for research and innovation:

> HMI [human–machine interaction] and driver information in EVs is the new frontier
> that automobile designers should have their hands on' J.Mays, Ford Vice President of
> Global Design. (Automotive Design 2010)

2.4 THE CHANGING NATURE OF THE DRIVING TASK

In the (distant) past, to operate a car a user had only to interact with the steering wheel and the pedals (once the engine was running); now the situation is quite different. Car driving is not only about mobility but comfort, enjoyment, and status (Walker et al. 2001a). Technology is rapidly changing in vehicles; hence information exchange between the driver, the vehicle, and the environment is of critical importance now more than ever before, especially when considering ever-increasing safety standards and user expectations (Harvey and Stanton 2013).

Of course, the growing complexity of in-car technology means interface design requires careful consideration, especially with the inclusion of in-car entertainment, satellite navigation, and various driver assistance systems (Harvey and Stanton 2013; Kujala and Saariluoma 2011); however, the complexity does not stop there. Nonconventional drivetrain vehicles, that is, hybrid electric vehicles (HEVs), plug-in hybrid electric vehicles (PHEVs), range extended electric vehicles (REEVs), and battery electric vehicles (BEVs), bring with them further layers of complexity; most involve more than one fuel system, and some are equipped with more than one drivetrain. Importantly, these new layers of complexity, and the human–machine interaction (HMI) issues they raise, have as yet received relatively little attention in the extant literature. Thus the challenge is to develop HMI design guidance that not only deals with the novelty and complexity inherent in modern, nonconventional drivetrain vehicles, but influences drivers to choose more energy-efficient driving behaviours.

2.5 DESIGN AND PERSUASION

Before discussing vehicle-specific HMI design a broad exploration of some general design philosophies is merited, in particular those design methods explicitly intending to influence behaviour.

The design of a technological product or system will influence users' perceptions of that system, and, as aforementioned, products are designed with specific activities in mind (Stanton and Baber 1998). To begin with, the designer of a technology must consider not only his or her own needs, but the needs of all potential users (Harvey et al. 2011a). Though this may sound relatively obvious it is important to bear in mind that people have a tendency to believe that their own needs and perceptions of a system are equally applicable to everyone else (Landauer 1997). It is also necessary to understand that technology design not only needs to be, but inherently *is* persuasive; this inherent persuasion, however, may not always be something which designers explicitly consider (Redström 2006), thus we need to acknowledge, explore, and understand it.

The acknowledgement that any technology design is necessarily persuasive, in that it guides (or persuades) a user to behave in a particular way with said technology, leads to the notion of intention. Fogg (2003) stated that intention is a characteristic feature of persuasion and that technology will *always* change what users think and do. Lockton et al. (2008a, 2008b, 2010) describe a similar philosophy in their discussion of the 'design with intent' method, an approach to design that explicitly recognises the intention to influence behaviour inherent in design. For example, a product's interaction means or sequence can be designed in such a way as to make users aware of their choices and the consequences of those choices; it is argued that this will have an effect on user behaviour. A simple example offered in Lockton et al. (2008a) is the two-button toilet flush used to bring water usage to the conscious attention of the user.

It is also possible to affect behaviour through goal setting and information provision. The aim here is to affect people's knowledge, beliefs, attitudes, and intentions; the determinants of behaviour (Ajzen 1991; Fishbein and Ajzen 1975). According to Abrahamse et al. (2005) energy use intervention strategies are more effective if they target these behavioural determinants. Abrahamse et al. (2005) argued that there are two primary types of behaviour intervention strategy relating to energy use; antecedent strategies and consequence strategies. Antecedent strategies encompass methods that involve providing an individual with information before the behaviour in question is performed. Consequence strategies involve punishing or rewarding certain behaviours *after* they have occurred. Feedback provision falls in the latter category. Individuals have perceived self-efficacy as a change in behaviour results in a change in subsequent feedback.

The effect of feedback (i.e. an indication of the consequences of a person's actions) on performance has long been recognised in the field of psychology (Ammons 1956; Bilodeau and Bilodeau 1961). It has been suggested, however, that knowledge of the consequences of behaviour is not a sufficient condition for effective performance; feedback and an individual's goals interact to steer performance (Erez 1977). In early work, Locke and colleagues found that the effect of feedback on performance is mediated by an individual's goals and intentions (Locke 1967, 1968; Locke and Bryan 1968, 1969a, 1969b). An interesting point to note here is that the goals driving behaviour do not necessarily have to be self-set; people provided with a goal that they had no part in developing still demonstrate energy conservation behaviours when supplied with feedback (Erez 1977; McCalley and Midden 2002).

Regardless of the reasoning behind, or source of an individual's energy saving goal, it is possible that the feedback itself prompts goal activation, without the need for explicitly drawing attention to the requirement for energy conservation. According to the feedback intervention theory (Kluger and DeNisi 1996) feedback directs an individual's attention to a goal, and a specific goal level. Goals can be described in terms of the different levels of behaviour to which they apply. For example, a person may have a high-level, over-arching goal of wanting to be eco-friendly. The goal of wanting to use less energy on a particular car journey is a low-level goal – it is specific to the task at hand. McCalley (2006) furthered the discussion arguing that goals *must* be specific and task related (rather than high level) in order to affect task-specific behaviour. For example, to reduce energy use while driving, activating the goal of 'I want to be eco-friendly' is not sufficient; a specific driving-related goal must be activated (McCalley 2006).

It is critical to understand the interconnectedness of goals and feedback if we are to take advantage of them in encouraging sustainable driving behaviour through design. According to the goal-setting theory (Locke and Latham 2002) a goal can only be effectively reached if appropriate feedback is provided such that the individual can know where they stand in relation to that specific goal (Locke and Latham 2002).

2.6   ENERGY USE BEHAVIOURS IN VEHICLES

Considering the aforementioned importance of user behaviour on energy usage (Lockton et al. 2008b; Zachrisson and Boks 2010), and given that this is particularly significant in the vehicle domain (Hawkins et al. 2013; Organisation of Economic Cooperation and Development 1993), it is important to look at how energy use in a driving situation can be affected by information provision and the activation of energy-related goals.

That a person's driving style can have a large effect on the energy use and emissions levels of the vehicle (Barkenbus 2010; Holmén and Niemeier 1998) is not a recently discovered effect; in 1979 Leonard Evans found that reducing acceleration levels and driving 'gently' in a real world setting resulted in a 14% fuel saving. This fuel saving was achieved without increasing trip time (Evans 1979). Similarly, Waters and Laker (1980) asked participants to drive 'economically' around a track on a second session of driving. After accounting for speed reductions (it seemed that some people assumed 'economical' equated to 'slow') a 15% fuel saving was demonstrated. Both of these studies demonstrated fuel savings using only the activation of a goal, that is, to use less energy, without the inclusion of feedback tools additional to the established driving environment (e.g. engine sounds, tachometer readings, perceptions of acceleration and deceleration).

In an early study by Hinton et al. (cited in Van der Voort et al. 2001) a driver support tool providing fuel use feedback was specifically examined; however, unlike the Evans, and Waters and Laker studies, only very small, insignificant fuel savings were brought about. The reason for this lack of effect was put down to inaccurate information that was often untimely, contradictory, and unclear (Van der Voort et al. 2001). Furthermore, the tools were considered to be distracting and were largely ignored. This highlights the fact that the presence of a driver support tool is not a sufficient condition for fuel conservation; the *design* of the tool must be carefully considered.

Designing a fuel-efficiency support tool requires attention to be paid not only to usability and aesthetics, but also to information content. Hooker (1988) found that gear shifting, speed choice, and acceleration and deceleration were the elements of driving behaviour that had the largest effect on fuel economy. Thus Van der Voort et al. (2001) investigated the efficacy of a prototype fuel-efficiency support tool that provided online feedback and advice to drivers based on these driving elements. Following on from the shortcomings of the Hinton et al. (cited in Van der Voort et al. 2001) study, Van der Voort and colleagues argued that a support tool must take into account the spatial and temporal context of the vehicle and must not be distracting. The support tool developed in the study was tested in a simulated environment with promising results. Participants provided with the tool and asked to drive as efficiently as possible achieved a 7% additional fuel saving over those participants without a feedback device (i.e. goal activation only). In a purely urban simulated environment this additional fuel saving rose to 14% (Van der Voort et al. 2001).

The studies presented thus far have all investigated fuel economy in ICE vehicles. While results from such research are highly informative it is necessary to look also at work in the hybrid and electric vehicle domain. For example, Bingham et al. (2012) highlighted the importance of driving style in electric vehicles. In this study the authors found that there can be as much as a ~30% difference in energy consumption between moderate and aggressive driving styles (Bingham et al. 2012). Moreover, as Kim et al. (2011) pointed out, range anxiety, and the lack of infrastructure and fast-charge options associated with electric vehicles mean drivers have a higher motivation to drive efficiently and to conserve as much energy as possible. In their study Kim et al. (2011) found that drivers presented with a visual representation of their acceleration behaviours (Figure 2.3) presented milder, more stable accelerator pedal usage and lower energy consumption than those without the feedback. This is of particular significance considering Cocron et al.'s finding that different driving styles have a much larger impact on fuel efficiency in vehicles with electric powertrains that in ICE vehicles (Cocron et al. 2011).

While Kim and colleagues were looking at power flows, Burgess et al. (2011) were investigating the option of displaying the number of miles left in the battery. In their simulator-based study, people were found to drive more economically *with* the display than *without* it (Burgess et al. 2011). This type of display also presents to the driver the added benefit in electric and hybrid vehicles of regenerative braking, the uptake of energy otherwise lost when applying the brakes. An issue here, however, is unfamiliarity; participants needed to adapt to the unfamiliar displays and to adopt a new style of braking. Furthermore the driving style is not the only

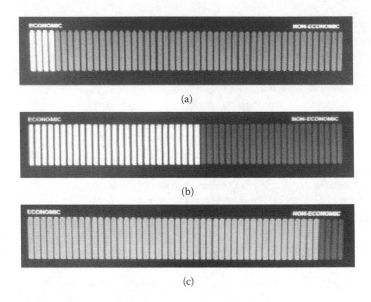

(a)

(b)

(c)

FIGURE 2.3. Power flow gauge investigated by Kim et al. 2011 (original in colour); (a) low acceleration, displayed to participants in green, (b) medium acceleration displayed to participants in amber, (c) high acceleration displayed to participants in red. (Adapted from Kim, S., et al., *Human-Computer Interaction, Part III, HCII 2011, LNCS 6763,* Springer-Verlag, Berlin Heidelberg, 2011.)

influence on the range of the battery; weather (particularly temperature) and road conditions also have large effects on battery performance (Burgess et al. 2011).

A further issue to consider is that of driver preferences: what type of guidance would people *want* to have, and how do they think it would affect their driving? For an individual to continually use a driver support system they must have a favourable opinion of it, otherwise they are liable to ignore it, or (if possible) switch it off. In a questionnaire-based study, Fricke and Sheißl (2010) found that respondents preferred the option of assistive visual information to that of direct intervention. An example of a direct intervention is the inclusion of resistance in the accelerator pedal to indicate overly rapid levels of acceleration; this was investigated in Larsson and Ericsson (2009), and although less rapid acceleration was encouraged, no significant reduction in fuel use was found. Accelerator-based haptic feedback was also investigated by Mulder et al. (2008); however, while improvements in car-following performance were found (in terms of safety), energy efficiency was not investigated (Adell and Várhelyi 2008). Whether one form of information is more effective at supporting economical driving than another, however, remains to be seen.

Although it has been suggested that drivers would welcome the introduction of pretrip, in-car, *and* posttrip eco-driving advice (Trommer and Höltl 2012), Stillwater and Kurani (2011) found that people with experience of the different tools prefer online, in-car feedback over offline, historical fuel use data (though this study did not investigate pretrip planning tools). Participants stated that in-car advice had more of an effect on their fuel use behaviours than offline information

(Stillwater and Kurani 2011); this finding can be explained using research showing that the closer, and the more often a reinforcement follows a behaviour, the stronger the stimulus–response relation becomes (Jager 2003). Lockton et al. (2008b), in their discussion of design with intent (their aforementioned design approach), also make this point; for behavioural adaption to be successful, feedback should be immediate.

Keeping drivers engaged in a system such that they will continue to use it and therefore show continued reductions in energy use can be partly achieved by considering subjective preferences like those outlined above; however, these may not be sufficient on their own. An adaptive system may provide a further means for maintaining eco-driving (see www.ecodrive.org) motivations. Wada et al. (2011) examined such a system. The feedback tool displayed to participants in the study responded to participants' behaviour inasmuch as the stringency with which economical driving was judged increased as drivers' eco-driving performance increased. Across 5 days of testing, participants with the adaptive tool achieved the highest energy savings compared to those with a nonadaptive tool, and to those without a tool (Wada et al. 2011). The authors argued that through adapting to drivers' skill the motivation for economical driving was maintained, resulting in continuous improvement in fuel economy. Participants were engaged as they could see themselves improving, an issue related to self-efficacy; they could maintain the challenge (Wada et al. 2011).

2.7 SAFETY AND USABILITY

An informative, aesthetically pleasing tool with which individuals are engaged and enjoy using may help to encourage efficient, environmentally-friendly driving styles, but that does not necessarily mean it will be appropriate for use in vehicles on the road. The practice of hypermiling (see www.hypermiler.co.uk) provides an interesting example of where a range of behaviours that have a significantly positive effect on energy conservation are not necessarily advisable due to safety reasons. Although overinflating tyres, turning off the engine and freewheeling downhill, and drafting as close as possible to the vehicle in front in order to make use of the slipstream may be beneficial activities for reducing fuel consumption, they present a trade-off in terms of road safety (Barkenbus 2010; Edmunds.com 2009).

The driving task is highly complex, comprising over 1600 separate tasks (Walker et al. 2001b). Being the safety-critical domain it is, the addition of more information to an already complex array of in-car systems should be very carefully considered if we are to avoid increasing workload and distraction, both of which are causal factors for accidents (Birrell and Young 2011; Pradhan et al. 2011). Take the Wada et al. (2011) study described above; although subjective workload ratings decreased with time in the control and nonadaptive display groups, those with the adaptive tool demonstrated higher workload scores. Importantly, these scores did not decrease with time. This may be problematic; people have limited cognitive resources, and as such, if the nondriving task demands increase (such as can happen when required to attend to an additional 'eco' display), attentional resources for other tasks may decrease (Wickens and Carswell 1997). This could result in the possibility that the concurrent feedback will interfere with ongoing task performance, a principle that has been demonstrated both within and outside the driving domain (Arroyo et al. 2006; Corbett and Anderson 2001; Stanton et al. 2011). Furthermore, Groeger (2000) describes driving as a goal-directed task, with multiple goals (e.g. speed, safety, economy) active simultaneously that at any point in time may be in conflict with each other. Highlighting the importance of economy goals may, therefore, have a detrimental effect on performance in other aspects of driving, for example safety.

Despite the possibility of conflict arising in the driving task, safe driving and economical driving do have significant overlaps (Young et al. 2011). Aggressive driving is seen as both dangerous (Young et al. 2011) and uneconomical (Ericsson 2001) due to characteristically high acceleration and deceleration rates, high engine speed, and power demands. It is possible then to encourage both safe *and* economical driving through supporting eco-driving; for example, Hedges and Moss (1996) showed that after supplying eco-training to Parcelforce van drivers accident rates dropped by 40% and fuel efficiency increased by 50%. Moreover, Haworth and Symmons (2001) demonstrated a 35% reduction in accident rates alongside reductions in fuel consumption (11%) and emission volumes (up to 50%) following similar training. Although these studies demonstrate some of the joint benefits of certain driving styles, they are both examples of antecedent strategies, that is they both employed pretask driver training, *not* concurrent feedback, thus they do not address the issue of distraction, a point noted by Haworth and Symmons (2001).

The distractive qualities of an in-car information system have been investigated by a number of researchers (Donmez et al. 2007; Harms and Patten 2003; Horberry et al. 2006, 2008; Lansdown et al. 2004; Reyes and Lee 2008), yet research primarily considering eco-feedback distraction effects is less abundant. As aforementioned, Wada et al. (2011) considered workload in their investigation of an adaptive cofeedback interface; however, this was relatively limited in its appraisal of distraction in that subjective workload scores were obtained only through questionnaires, not direct measurements of distraction. A study by Birrell and Young (2011; see also Young and Birrell 2012) did directly assess the impact on both fuel use and safety in an investigation of two versions of a smart driving tool, that is, a device that offers advice both on eco-driving matters (e.g. acceleration and deceleration rapidity) and on safety (e.g. lane departure, headway information). They found that participants with access to in-vehicle feedback displayed fewer speeding behaviours and fewer instances of aggressive acceleration and braking, beneficial for both safety and economy. Furthermore, drivers with the in-vehicle feedback also exhibited safer headway maintenance behaviours. These results were all obtained without significant increases in driver distraction. When investigating the efficacy with which participants performed a peripheral detection task while driving, Birrell and Young (2011) found that those with one of the two in-vehicle feedback systems investigated performed significantly better in an urban driving scenario, with no significant differences in other scenarios or with the other interface design. That the researchers examined two different interfaces again highlights the importance of the way in which information is presented; not only was one design superior in terms of the peripheral detection task results, that same design received significantly lower subjective workload ratings (Birrell and Young 2011).

It is clear that the way in which an interface is designed can have huge implications on its ability to elicit target behaviours, its acceptance by users, and its propensity to cause distraction and confusion. Results from Van der Voort et al.'s (2001) study, described above, led the authors to describe a set of user requirements for a fuel-efficiency support tool:

- Clear, accurate, non-contradictory information
- Account for the context in which the car is situated
- Not interfere with the driving task
- Work in urban and nonurban environments

Similar sentiments were put forward by Harvey et al. (2011a) for the design of in-vehicle information systems (IVIS). For such systems one of the main priorities must be to minimise conflicts with the primary driving task, thus reducing the likelihood of distraction. When designing such a system complexity is a major issue; that the driving context is highly complex necessarily means designing for usability in the driving context will be complex (Fastrez and Haué 2008). As such, the usability of an in-vehicle system must be defined specifically for the context of use (Harvey and Stanton 2012; Harvey et al. 2011a), and to test such a system requires repeated usability evaluations at different stages of the development process (Mitsopoulos-Rubens et al. 2011), with a variety of evaluation methods for example focus groups, user tests, and expert evaluations, including both subjective and objective usability measures (Harvey et al. 2011a; Tango and Montanari 2006).

2.8   ERGONOMICS AND THE DESIGN OF LOW-CARBON VEHICLE HMIs

The discussion up to this point has covered a variety of related topics, including CO_2 emissions induced by the use of fossil fuels, energy consumption and conservation, persuasive design, behavioural change, user preferences, distraction, and usability. Knowledge of these related elements provides the basis for the ongoing aims of this review chapter, and allows for the suggestion of where researchers might focus their efforts to have most beneficial impact on the issue of sustainability in private transport. Bringing these topics together, it is possible to see more specifically where lie some potential future challenges in low-carbon vehicle interface design, or indeed any in-vehicle information design (as shall be discussed), and how the ergonomics and design communities could meet these challenges.

Four primary areas have been identified as offering potential for the beneficial application of ergonomics to the design of the in-vehicle environment; the necessity to overcome the significant and oft-cited issue of range anxiety (Cocron et al. 2011); the need to support the development of accurate mental models of the novel, often poorly understood technology; the issue of rising in-car complexity, and the effect this will have on workload, distraction, and the resulting safety implications; and the opportunity to take advantage of this novelty in fostering the development of new, economical, yet safe driving habits. Although these four concerns have been stated separately, it should be noted that they are interrelated, inasmuch as any single design intervention strategy will likely need to be considered in terms of its impact on all four issues. It is for this reason that ecological interface design (Burns and Hajdukiewicz 2004; Vicente and Rasmussen 1992), a design method that considers the system in its entirety taking into account the interrelatedness of system components and functions, was initially chosen as a method to address such an issue.

2.8.1 Ecological Interface Design

Ecological interface design (EID) is based on the tenets of cognitive work analysis (CWA), a formative analysis technique that describes how a system *could* perform given the constraints of the domain and the functional links between low-level system components and high-level system functions and purposes (e.g. Jenkins et al. 2009; Rasmussen et al. 1994; Vicente 1999). The technique is posited as applicable to first-of-a-kind systems for which there are no precedents (Vicente 1999) and as such was considered to be aptly poised as a basis for developing a driving feedback tool for use in electric and hybrid vehicles. EID is essentially about representing the environmental constraints, or boundaries (graphically or otherwise), of the domain such that direct perception is possible, thus removing the requirement for indirect mental representations of external reality. Creating and maintaining an indirect representation of the world is problematic in that not only does it require more cognitive resources to construct (particularly significant considering the safety-critical, cognitively demanding nature of the driving task), but also is more susceptible to inaccuracies (Gibson 1979), with such inaccuracies leading to an incomplete and incorrect understanding of the system or environment in question. Though CWA and EID are more often applied to larger, more complex sociotechnical systems than the interface of a single vehicle, for example in nuclear power (Olsson and Lee 1994), the military (McIlroy and Stanton 2011) and aviation (Burns and Hajdukiewicz 2004), there are a number of examples where the design methodology has been used in vehicle design; these will be introduced as the discussion of the four challenges progresses.

2.8.2 Overcoming Range Anxiety

Range anxiety, arguably the most influential of barriers to electric vehicle uptake (Pearre et al. 2011), has been shown to decrease with experience in an electric car (Burgess et al. 2011; Cocron et al. 2011; Franke et al. 2011; Krems et al. 2010). Through design it may be possible to further reduce, even eliminate range anxiety, as well as speed up the time with which the anxiety wanes. Turrentine et al. (2011) and Pearre et al. (2011) argued that a safety margin of around 20 miles is required to alleviate range anxiety (a 'range buffer'); however, Franke et al. (2011) argued that it could be possible to overcome range anxiety with information and interface design (see also Cocron et al. 2011). Despite finding suboptimal range utilisation in their field study of electric vehicle drivers (i.e. range buffers were indeed used) they put forward the argument that increasing the range of an electric vehicle may be less important than merely providing the driver with reliable, accurate information about the usable range of the vehicle. Importantly, it is about reducing the *perceived* barriers associated with range anxiety (Franke et al. 2011). This allows for the suggestion that range anxiety could be reduced (eliminating the requirement for range buffers) if the car–driver–environment interface is sufficiently well designed, in both information content and presentation. Though discussions explicitly linking EID and range anxiety are not, to our knowledge, available in the extant literature, it is in a driver's (mis)understanding of the system in its entirety (including the vehicle, the driver, and the environment in which they find themselves) that range anxiety partly finds it basis; this is intimately linked with how a system is represented, and the resulting mental models developed and maintained by the user.

This line of thought can also be applied to the act of driving itself. Research on driving behaviour and efficiency suggests that the average driver could save around 10% of the fuel they use for a given journey simply by changing their driving style (Barkenbus 2010). Additionally, and as aforementioned, Bingham et al. (2012) suggested that the difference in energy consumption between moderate and aggressive use of an electric vehicle could be as high as 30%. This relates, in part, to the vehicle's regenerative braking capabilities; these will only work optimally with smooth deceleration profiles (i.e. avoiding harsh braking, in which the mechanical brakes are employed thereby bypassing the regenerative braking mechanism). To help the driver increase their range (and alleviate range anxiety) this then becomes a question of helping the driver to use their vehicle in the most efficient way possible, that is, to drive economically. Importantly, though Bingham et al. (2012) used the electric vehicle as a platform for their research, their results were argued to also be applicable to plug-in hybrids. Given the wide variety of research, spanning almost 40 years, into fuel efficiency in the conventional ICE vehicle (Evans 1979; Staubach et al. 2014b), we would argue that this is applicable to *all* private road vehicles. Although increasing range may be particularly important in vehicles with reduced range capabilities, decreasing fuel consumption is important *regardless* of vehicle type. How to present such information to the driver is something with which EID, and its underlying theoretical foundations, may be able to help us.

2.8.3 SUPPORTING ACCURATE MENTAL MODELS

The assertion made by Franke et al. (2011) on the importance of overcoming *perceived* barriers implies that the barriers are not necessarily present in the physical world but are based in people's beliefs, right or wrong, about electric vehicles and the range they are likely to require. The question of how to design to overcome barriers then becomes a question of how to represent the car environment system to the driver such that they are fully aware of all the parameters, that is, it is about supporting an accurate mental model of the system (Gentner and Stevens 1983). This is also true for driving behaviour itself; in a more recent survey of hybrid electric vehicle drivers, Franke et al. (2016) found that respondents had many different conceptualisations of energy efficiency in the vehicle, including a number of false beliefs that served to impair drivers' efforts to use their vehicle efficiently. An in-vehicle system designed in such a way that the actual energy use characteristics of the vehicle are presented, or in a way that displays to the driver the most efficient way in which to use that vehicle, may allow the driver to develop an accurate mental model. Subsequent false beliefs would then be less likely to arise.

This arises from the idea that when a user does not have an accurate or sufficiently detailed understanding of a system (i.e. they lack an accurate mental model) undesirable behaviour is more likely (though see Revell and Stanton (2012) for an in-depth discussion of mental models). Using Lewis and Norman's (1986; see also Reason 1990) terminology, this is about mistakes as opposed to slips; a slip is where a user intends to perform the desirable action, but performs it incorrectly, whereas a mistake is where a user intends to perform an undesirable action. The defining difference is intention; slips are unintentional, but with mistakes the *action* is intentional; the user simply does not know the action is incorrect or undesirable.

For example, Franke et al.'s (2011) participants may have displayed suboptimal range utilisation due to their incomplete or incorrect mental models of the system. It can also be argued that respondents to their more recent study (Franke et al. 2016) reported false beliefs about efficient use of their vehicle due to incorrect mental models. Although the respondents reported beliefs and behaviours that they thought to be good for fuel efficiency, some were, in fact, *detrimental* to fuel efficiency. For example, one such behaviour reported was the maximisation of the use of the electric motor over the combustion engine, based on the belief that the electric motor is more efficient. This is not necessarily the most efficient strategy (see Franke et al. 2016); to perform such a behaviour, thinking that it represents the most efficient strategy, could therefore be described as a *mistake* in Norman and Lewis's terminology.

2.8.4 WORKLOAD AND DISTRACTION

Of course, any in-vehicle information system or interface must be considered in terms of its impact on workload and distraction. For example, although adaptive cruise control (ACC) is aimed at reducing the workload of the driver, if the way in which it functions is not wholly apparent (i.e. the interface is not sufficiently well designed) then the issue of mode error can result, that is to say, the user does not understand in which mode the automation is functioning, or how or why the automation is functioning in the way it is (Liu et al. 2006). The resulting confusion could wholly undermine the intended benefits of the system.

Seppelt and Lee (2007) investigated the use of EID in the development of a visual representation ACC, finding that an EID-informed display supports safer driving behaviours when the ACC was activated and when driving manually, leading the authors to argue that providing drivers with information regarding the state of the automation was more useful than simply providing collision warning alerts (Seppelt and Lee 2007). Similarly, Mendoza et al. (2011; see also Lindgren et al. 2009) applied EID to the design of an advanced driver assistance system that provided staged warnings relating to a number of safety systems, with results from the simulator study suggesting that EID can offer safety benefits, particularly in terms of lateral position and distance to the lead vehicle. Such staged warnings represent the display of system boundaries (i.e. the boundaries between safe and unsafe operation), a key principle of EID. In this last study it was pointed out that a potential source of distraction is the presentation of information not relevant to the situation. As Kaufmann et al. (2008) argued, there may be a risk of driver distraction from unimportant information presentation during safety-critical events.

That EID can specify *what* an interface has to display in a given situation or for a given function (Lee et al. 2004), through the preceding analysis (using CWA) of the functional links between lower-level system components and higher-level system functions, has led to the suggestion that EID can help design interfaces that avoid the problem of displaying irrelevant information (see Chapter 5 for a more detailed discussion on EID's contributions to design). Young and Birrell's studies outlined above (Birrell and Young 2011; Young and Birrell 2012) highlight this; the smart driving tool judged to be superior in terms of workload and distraction was designed using the principles of EID, with the design process following a work domain analysis (an analysis step integral to both EID and CWA) of the driving domain (Birrell et al. 2012). Other such studies include that of Jenkins et al. (2007), who developed a lateral collision warning system using this analysis step (i.e. work domain analysis), finding that it compared favourably with existing lane departure warning systems. Also, Lee et al. (2006; see also Stoner et al. 2003) tested an interface developed using CWA and EID for the support of manoeuvres requiring a lane change, with results showing that EID-inspired displays performed at least as well, if not better than traditional displays, particularly in situations where the participants could only view the scenario for a short period of time (Lee et al. 2006). The ability of an interface to be understood quickly is an important feature if it is to avoid being distractive. Making the boundaries clear to the user, for example the boundary between safe and unsafe operation (as aforementioned), or indeed the boundary between efficient and inefficient driving, therefore becomes of particular interest. Using the theory behind EID may help to achieve this aim without also incurring additional workload or distraction.

Research into the effects of EID interfaces on workload is not, however, clear cut. Stanton et al. (2011) investigated an interface displaying the functioning of stop and go adaptive cruise control (S&G-ACC; an extension of ACC that includes operation at slow speeds and over short distances). Though in this research EID was not specifically mentioned, the design of one of the three interfaces under examination was in line with the principles of EID insofar as it directly represented the radar capability of the technology. The study paid particular attention to workload, citing earlier research that suggested there is an *increase* in workload associated with monitoring the activities of an automated system (Stanton and Young 2000, 2005). It was demonstrated that providing a direct representation of system state, rather than simply providing warnings regarding new vehicles entering the following path, allowed for a fuller understanding of the operation of the automation, thereby supporting safer driving behaviours (Stanton et al. 2011). The results also showed, however, that the more detailed radar-type display (in line with EID tenets) incurred higher workloads.

In terms of encouraging *efficient* use of the vehicle, rather than simply *safe* use, it is useful to return to the issue of displaying the correct type of information, and displaying system boundaries. As evidenced in Franke et al.'s (2016) study of hybrid electric vehicle drivers, even experienced users hold false beliefs. The authors make the point that the hybrid vehicle is particularly complex, involving two fuel systems and two drivetrains that interact with one another in different ways, depending on the characteristics of the driver's behaviour, on the route driven, and on the current energy reserves in the fuel tank and batteries. To display such information to the driver would likely require a relatively complex, visual display. Although EID may be well suited to designing such a display, it may not necessarily do so in a way that reduces workload (Stanton et al. 2011). Furthermore, the in-vehicle environment is already replete with visual information; the addition of further visual systems should be carefully considered. Hence rather than attempting to apply a full EID process to the design of in-vehicle information, it may be equally useful (in terms of encouraging efficient driving behaviours, and therefore efficient use of the vehicle), yet incur lower workload, to focus simply on displaying to the driver the boundary between efficient and inefficient driving styles, that is, supporting efficient use.

Though this is a departure from the full EID process (as will be discussed in more detail in Chapters 5 and 6), it is based on one of the method's key foundational principles, that is, to display system boundaries, or constraints. Indeed, it is this very point that has guided the research presented in the latter stages of this book (see Chapter 6, e.g.), and is a point to which Franke et al.'s participants alluded (Franke et al. 2016). When asked about the kind of support systems drivers would like in the vehicle, respondents provided a number of suggestions relating to the location of particular system state:

> ... drivers suggested that certain critical system states should be more clearly displayed (e.g. the point of maximum efficiency of the combustion engine, the neutral point at which there is zero energy flow in the system, the point at which regenerative braking is optimal, or a point just before that at which the combustion engine turns on), and that targeting these points should be facilitated. (Franke et al. 2016, 39)

Such descriptions of the point of maximum efficiency, the point of zero energy flow, the point at which regenerative braking is optimal, and the point at which the combustion engine turns on, can be conceptualised as boundaries between certain system states, or boundaries between efficient and inefficient use of the vehicle. Supporting perception of these boundaries may not only help increase driving efficiency, but may circumvent some of the potential workload and distraction issues that could be brought about by attempting to add further, complex visual display systems to the vehicle.

2.8.5   DEALING WITH COMPLEXITY AND TAKING ADVANTAGE OF NOVELTY

The studies outlined above suggest that EID, and its underlying theoretical principles (see Chapters 5 and 6 for more detail), could bring significant benefits to the driving domain. As part of an ongoing, multiple partner research project, Young and Birrell have successfully brought together safety and economy advice in one EID-guided information tool (Birrell and Young 2011; Birrell et al. 2012; Young and Birrell 2012); interestingly, the tool that was developed not only attempted to display system functioning, but actually informed the driver of particular behaviours that could be executed, at particular times, in order to maximise safety and efficiency. Although the tool was designed with a traditional ICE-powered vehicle in mind, many of the behaviours that characterise efficient driving (e.g. smooth accelerations, anticipation, avoidance of harsh braking events) are applicable to vehicles of any fuel system.

There are, however, additional considerations that should be made when designing for vehicles with unconventional drivetrains. For example, products that people perceive to be 'eco-friendly' can incur excessive use; the 'rebound effect' describes how a product is used more often if a user thinks each use is less environmentally damaging (Berkhout et al. 2000). The extra usage incurred negates any improvements in the energy savings made through the design of the product. This is particularly important for electric cars; just because tailpipe emissions are zero doesn't mean the electricity required is clean and abundant. Is it possible, therefore, to develop an interface that discourages this kind of behaviour? Again, this harks back to a central aim of the research presented in this book, that is, how can we develop an in-vehicle interface that supports efficient use of the vehicle?

As aforementioned, in most road vehicles the vast majority of in-vehicle devices that aim to help drivers increase their efficiency simply display to the driver the current energy or fuel consumption rates (Wellings et al. 2011). Regarding this question, the metric used for displaying driving efficiency is an issue that must be addressed. Traditionally, miles-per-gallon has been used throughout the industry (in the UK and US); however, this will clearly not suffice for vehicles incorporating batteries into the power system. Such issues were considered by Stillwater (2011); he argued that miles-per-gallon can, in many situations, be misleading, insofar as it only provides a 'tank' metric, not a real-time energy balance. Simply offering the number of miles left in the battery (or battery/fuel tank combination) may not be advisable either; such a display, if it were to be accurate and consistent (Van der Voort et al. 2001), would have to consider not only the effect of the weather (most importantly temperature) on the battery, but the topography of the route and the effects of regenerative braking (Stillwater 2011). Perhaps, therefore, it may be more suitable not to display to the driver their current energy use statistics, or power flow information, but to provide information that actually guides the fuel-efficient behaviours themselves. Once again, can we display the boundary between efficient and inefficient use of the vehicle using in-vehicle information?

Tackling the HMI challenges posed by the widespread uptake of electric and hybrid vehicles, in safety, economic, and enjoyment terms, will require careful consideration; however, it is important not to lose sight of the opportunities provided by such a technological advancement. As has been described, there is a large potential for environmental benefit arising from encouraging behavioural change through design. With a novel product (such as a low-carbon vehicle), taking the opportunity to foster behavioural change can have long-lasting results. Zachrisson and Boks (2010) argued that a product's ability to break old habits is related to the novelty of the interaction with that product, with more innovation or novelty having a stronger ability to break previous habits. This may be because prior schemata are evoked to a far lesser extent when interacting with a novel product than when interacting with a more familiar product. Schemata can be conceptualised as organised knowledge structures, based upon past experiences, that interact with information in the external environment to guide behaviour in a given situation (Bartlett 1932;

Plant and Stanton 2012; Stanton and Stammers 2008). According to Neisser (1976) existing schemata affect the way we perceive the world, influence the decisions we make, and direct our actions. If the situation or environment is one of novelty then it is unlikely that a fully developed schema will exist to guide behaviour.

When considering this in a driving context, the more familiar the human–vehicle interaction or interface design, the more similar to prior driving habits the observable behaviour will be, including any previously learned bad habits. A novel interaction will more readily support the modification (Piaget 1952) of previously held schemata. Importantly, schemata are active (Neisser 1976) and as such, if the user has a positive behaviour that would be beneficial to turn into a habit, the product should maintain the context around the behaviour as stable as possible, thus helping to develop a more economically-framed driving schema. Hence it is argued that interface design in electric and hybrid cars, aided by the novelty of the technology and its ability to encourage schema adaption and development, can be used to foster new economical driving styles, replacing fuel-intensive habits.

2.9 CONCLUSIONS

Considering the backdrop of the overuse of energy resources and the excessive production of waste and emissions, alongside the introduction of new vehicle technologies, it is clear to see that ergonomists are faced with potentially challenging, yet promising opportunities with regard to the design of the HMI in the private road vehicle. The need to present additional information, that is, fuel use and economy information, and the need to represent the more complex nature of the low-carbon vehicle (particularly for hybrid vehicles, which have more than one fuel type), both have implications for workload and distraction, and hence safety. However, it is important not to lose sight of the prospects for encouraging behaviour change through in-vehicle information design. With careful interface design it may be possible not only to help an individual reduce their energy consumption, but also to alleviate the problem of range anxiety through supporting them in their maximisation of their vehicle's energy reserves.

Although the preceding discussions have focused predominantly on low-carbon vehicles (in terms of the necessity to make the most of the limited range inherent to electric vehicles), the benefits of supporting efficient driving behaviours are not limited to such technologies. Efficient use of the vehicle is still recognised as of particular importance when discussing cars with limited range (i.e. electric vehicles); however, the remainder of this book takes a broader approach, focusing on how to help drivers make the most out of their energy reserves *in any vehicle*. This book as a whole therefore focuses on eco-driving more generally.

Eco-driving is the term that encompasses the behaviours that characterise efficient use of the vehicle, and it is this practice that provides the sole focus of the following chapter, in particular the general public's perceptions of it. As will be seen, the majority of this book is concerned with the provision of information to the driver, via an in-vehicle information system, to support and encourage efficient driving behaviours. However, it is important to recognise that this is not the only method by which efficient driving can be encouraged, nor is it the only potentially interesting avenue for research. The following chapter therefore takes a step back from looking specifically at *encouraging* specific fuel-efficient driving behaviours to look more generally at the practice itself, the perceptions that general public have of it, and at the levels of knowledge people already have of the specific strategies available for fuel conservation in the vehicle.

3 Driving and the Environment

An Exploratory Survey Study

3.1 INTRODUCTION

In the previous chapter a broad review of the literature pertaining to the effect of design on behaviour was offered, as well as an introduction to some of the challenges faced when considering the burgeoning interest in electric vehicles. Ecological interface design (EID) was also introduced, and the framework discussed in terms of its potential to guide the design of an in-vehicle information system that could help overcome some of these challenges.

As has been discussed, the original focus of this research project was the encouragement of the uptake and efficient use of the low-carbon vehicle. The previous chapter went some way to discuss the benefit of simply supporting efficient use of the vehicle, and how this is likely to be of particular importance in vehicles with limited range (i.e. electric vehicles). To encourage uptake is something that will require a wide variety of measures, from pricing to policy, to education, to training design. These aspects do not provide the focus of the remaining chapters of this book; rather attention is targeted towards the *use* phase of the vehicle. Although this directs focus, somewhat simplifying the research effort (compared to an attempt to address all the broad challenges outlined in the previous chapter), the challenge is still significant. As with encouraging low-carbon vehicle uptake, encouraging the adoption of eco-driving techniques is something that will also require a variety of measures. However, the benefits of doing so will be felt not only by those few early adopters of hybrid and electric vehicles, but will be seen by all drivers, regardless of the fuel system used by their vehicle.

In the previous chapter the concept of eco-driving itself was introduced, but not discussed at length. This chapter therefore specifically addresses the topic. As aforementioned, the focus is not on in-vehicle information specifically; rather it is on the practice of eco-driving from a broader perspective. Before embarking on an attempt to support such behaviours in the vehicle, it is important to recognise that there are a number of interesting and potentially fruitful avenues for research when considering the encouragement of any sustainable behaviour. As with home energy use (see Chapter 1), information provision at the point of use is only one such method for the encouragement of energy conservation behaviours (Abrahamse et al. 2005). Moreover, people's previously held knowledge and attitudes are important when attempting to encourage such practices. This chapter therefore investigates these concepts from an eco-driving perspective.

3.2   KNOWLEDGE OF AND ATTITUDES TOWARD ECO-DRIVING

When one searches the term 'eco-driving' using Google, the following quote (taken from www.ecodrive.org) appears at the top of the page:

> Ecodriving is a term used to describe energy efficient use of vehicles. It is a great and easy way to reduce fuel consumption from road transport so that less fuel is used to travel the same distance.

The first sentence of this perhaps slightly vague definition does in fact sum up the essence of eco-driving quite well, but says nothing of how it is to be achieved. As was briefly discussed in the previous chapter, eco-driving is a term encompassing the behaviours in the vehicle that characterise efficient use, for example smooth acceleration, maintenance of speed at low engine revolutions (i.e. RPM), anticipation of the road ahead, early gear changes, properly inflated tyres, and consideration of additional energy uses (e.g. air conditioning; the 'golden rules of eco-driving' according to www.ecodrive.org). It is all well and good to have academics provide research evidence for its benefits, and discuss at conferences the potential global energy (and emissions) savings brought about by eco-driving; however, if members of the general public do not know how to do it, are unaware of it or have an unfavourable opinion of it, it will remain a practice confined to the pages of academic publications and online motoring forums. This chapter, therefore, addresses the question: what do people know of eco-driving, and what do they think of it?

The previous chapter of this book provided a number of statistics pertaining to anthropometrically exacerbated (if not caused) climate change, with the point made that it is now generally accepted within the academic community that we, as humans on this planet, are using resources (and emitting waste products) in an unsustainable manner (Intergovernmental Panel on Climate Change 2007). A question that has not thus far been addressed is whether or not the public at large are aware of and agree with the academic community on these issues.

The popular news media, that is to say television news, daily broadsheet and tabloid newspapers, and popular news websites, significantly influence the public's understanding of climate science and policy (Wilson 1995). Biases abound in the media at large, and some outlets differentiate themselves by offering increasingly more opinionated content, regardless of scientific basis (Hmielowski et al. 2013). For example, in the United States, Fox News has been criticised for its dismissive attitude towards climate change (Feldman et al. 2011). This network has traditionally been of a conservative leaning nature; however, in the UK even the BBC, a publicly funded organisation founded with notions of impartiality, that has since been argued to be both left-leaning and right-leaning (Berry 2013), has had called into question the airtime they have given to deniers of climate change (Knapton 2014). On the other hand, there are myriad news outlets that present information far more in line with the academic community's consensus, including those that present a more liberal, environmentally focused standpoint, the most commonly known (in the UK) being *The Guardian* (see for example Hulme and Turnpenny, 2004; though we of course accept that all media outlets have their own biases).

Given this backdrop of contradictory information and opposing biases it might be expected that public opinion may not be in line with scientific consensus (i.e. approximately 97% agreement) that humans are causing climate change (Cook et al. 2013). Research suggests that this is indeed the case, in the United States (McCright et al. 2013), in China (Yu et al. 2013), in Europe (Engels et al. 2013; Poortinga et al. 2011), and elsewhere (Vignola et al. 2013). But what about eco-driving? Climate change is a well-known, commonly discussed topic; eco-driving is not. Do public perceptions of eco-driving marry up with the views of the scientific community?

It is useful here to reiterate some of the arguments made in the previous chapter, with regard to the effect of driving style on fuel use. As has been previously discussed, the effect is now relatively well known; more than 30 years ago Evans (1979), and Waters and Laker (1980) demonstrated around a 15% fuel saving, while more recently Barkenbus (2010) suggested 10% as an average figure that, with training and feedback, could be a sustainable saving. Similar results can be found elsewhere, for example: Wu et al. (2011) showed 12%–31% differences in fuel use with differing acceleration and deceleration behaviours; FIAT found that, with in-vehicle information, the average person saves 6%, with the top 10% of 'eco-drivers' saving 16% (Fiat 2010); Bingham et al. (2012) showed an energy use difference of up to 30% arising from the way in which an electric vehicle is driven; Van der Voort et al. (2001) showed 9% fuel savings by simply asking people to drive economically, and 16% when also providing participants with eco-feedback; and Gonder et al. (2011) suggested as much as 30%–60% fuel savings could be realised with extreme drive cycle differences (though even moderate driving styles could be improved upon by 5%–10% simply through driving behaviour).

Research concerning the knowledge the public at large have of eco-driving is less abundant, and such an understanding would help academia, government and industry to develop means for supporting such behaviours, be that through in-vehicle devices, policy interventions or educational strategies. This exploratory, survey-based research therefore addresses this knowledge gap.

In terms of existing literature addressing this issue, three such studies stand out; Delhomme et al. (2013), in an analysis of the self-reported frequency and difficulty with which people exhibit fuel-efficient driving behaviours; Harvey et al. (2013), in an investigation of the fit between eco-driving attitudes and environmental attitudes more generally; and King (2011), in a study of people's pre-existing knowledge of specific eco-driving strategies.

In Delhomme et al. (2013) respondents to a survey reported anticipation to be the easiest and most frequently adopted eco-driving strategy, and early gear changes and low motor revolution maintenance to be the most difficult. It was also found that those with higher environmental concern, and older drivers (particularly older females) reported lower difficulty in adopting fuel-efficient driving behaviours, and a higher frequency of performing them.

Where Delhomme and colleagues looked at the link between environmental concern and the reported performance of eco-driving *behaviours*, Harvey et al. (2013) investigated the relationships between *attitudes*. In a survey of 350 respondents, the authors found *no* strong links between price, convenience, environmental attitudes or attitudes towards eco-driving. In contrast to Delhomme et al.'s finding, Harvey et al. found that environmental concern is not of high enough priority to affect driving behaviour, and that eco-driving is considered less important than convenience. Although respondents reported high concern for the environment, this concern was not reflected in self-reported behaviours. Moreover, though in-vehicle feedback was considered as having potential to encourage the uptake of eco-driving behaviours, they were argued to be insufficient on their own.

Finally, King's (2011) report of a survey of New Zealand Automobile Association (AA) members provides us with information regarding people's *knowledge* of eco-driving behaviours. In addition to finding that most drivers already think that they are better at eco-driving than the 'average' driver (Alicke et al. 1995), it was found that older drivers were more able to provide eco-driving tips (i.e. descriptions of specific in-vehicle eco-driving behaviours), and that males were able to provide more valid tips than females. The most commonly reported behavioural strategy related to light acceleration and braking; however, the survey also included behaviours performed outside the vehicle (e.g. maintenance, weight reduction, tyre inflation) despite asking specifically for fuel-saving behaviours performed *while driving*. Moreover, tyre inflation was mentioned in the survey *before* asking respondents to provide fuel-saving tips; this very tip was the second most commonly reported. Knowledge of eco-driving was not, however, particularly high, with only around half of respondents providing two or more distinct fuel-saving tips, and only 50% of respondents referring to light acceleration and braking as good for fuel economy. Self-reported propensity to *perform* those behaviours was also investigated; over 5% reported never following their own advice, and around 20% reporting doing so only sometimes.

The AA survey also looked at attitudes towards eco-driving, finding that although it is generally popular (with 88% support), people were more interested in learning defensive driving, with fewer than 5% of respondents being prepared to invest a realistic sum in driver training (King 2011).

These three investigations provided the starting point for the current research, namely the investigation of the general public's general perceptions of eco-driving, of their knowledge of specific eco-driving strategies, and of the relationships between general environmental attitudes, eco-driving knowledge and the self-reported propensity to perform eco-driving behaviours.

3.2.1 Perceptions and Self-reported Ability

The first research question is an open one; what perceptions do people have of eco-driving? This question will be addressed in terms of two primary sub-questions: (1) whether people think it is a good idea in general and (2) the potential fuel savings that could be achieved by themselves and by the 'average' driver. This second question replicates that reported by King (2011), and relates to the finding that people consider themselves to already be more efficient than the 'average' driver. Such over-confidence in one's own ability is a staple finding in psychology (Alicke 1985; Alicke et al. 1995), and has previously been demonstrated in the driving domain (McKenna et al. 1991; Svenson 1981).

Whether or not estimates provided by the general public reflect those seen in the academic literature (Barkenbus 2010) is also of interest. If people consider the effect to be insignificantly small, they may be less likely to think it worthwhile; however, if they assume the effect to be greater than is likely to be the case, they may become frustrated with their (perceived) lack of success and give up the practice altogether.

3.2.2 KNOWLEDGE, ATTITUDES AND BEHAVIOUR

The second research question is also exploratory, again reflecting King's (2011) line of enquiry: what do people know of eco-driving, and are they aware of their levels of knowledge? Specifically, when asked to provide eco-driving tips, which behavioural strategies are most commonly reported, how many different behavioural strategies do people report, and do the reported strategies correspond to those present in the literature?

Relatedly, does self-reported knowledge correspond to actual knowledge? In other words, do people's self-rated knowledge of eco-driving on a scale (i.e. from 'never heard of it' to 'confident how to do it') correspond to a quantitative measure of eco-driving knowledge? Early research from Lichtenstein and Fischhoff (1977) indicates that people are generally good judges of their own knowledge, but that judgements are prone to biases of overconfidence. Although it is difficult to provide *absolute* measures of eco-driving knowledge (Cronbach 1975) it is possible to measure *relative* knowledge of eco-driving, and investigate whether those *relatively* less knowledgeable also provide lower self-assessments of their own knowledge than those relatively *more* knowledgeable of the practice.

The third research question concerns the possible link between attitudes and knowledge, asking whether or not those with more pro-environmental attitudes also hold more knowledge of specific eco-driving behaviours. A consistent and positive (although relatively weak) relationship between environmental knowledge and attitudes has been found in the past (Arcury 1990; Flamm 2006). Does this extend to eco-driving knowledge? Specifically, are those with more pro-environmental attitudes also more knowledgeable in terms of specific fuel-saving, in-vehicle behaviours (i.e. able to provide more valid eco-driving tips)? Following Arcury and Flamm, it is hypothesised that those with more pro-environmental attitudes will also be more knowledgeable of eco-driving behaviours.

The fourth research question concerns the relationship between attitudes and behaviours. It has been demonstrated that those with more pro-environmental attitudes are also likely to report engaging in more pro-environmental behaviours (Bamberg and Möser 2007); does this extend to eco-driving behaviours? Specifically, do those with more pro-environmental attitudes also report a higher propensity to perform specific in-vehicle, eco-driving behaviours? Regarding this, it is hypothesised that those with more pro-environmental attitudes will also report a higher propensity to perform eco-driving behaviours.

The fifth research question considers the relationship between knowledge and behaviour. Greater knowledge of potential action strategies has been shown to be associated with higher self-reported performance of pro-environmental behaviour (Hines et al. 1987); again, does this extend to eco-driving specifically? That is to say, do those people more knowledgeable of the means for driving efficiently (i.e. able to provide more valid eco-driving tips) also report performing those behaviours? Given

Hines et al.'s finding it is hypothesised that those who are more knowledgeable of eco-driving behaviours will also report a higher propensity to perform them.

3.2.3   GENDER, AGE AND EDUCATION

As aforementioned, in Delhomme et al. (2013) older, female drivers were found to report a higher likelihood of adopting eco-driving behaviours; however, in King (2011), it was demonstrated that older, male drivers held more knowledge of eco-driving behaviours. Research question 6 therefore addresses gender and age differences. With regard to gender, the question is left open; however, given the two results described above, it is hypothesised that older drivers will hold more knowledge of eco-driving behaviours than younger drivers, and will report a higher propensity to perform those behaviours.

In terms of education, Diamantopoulos et al. (2003) found that individuals with higher levels of general education also had higher levels of environmental knowledge, particularly for those with degree-level qualifications (e.g. BA, BSc). Hence the seventh research question asks if this relationship is also true for eco-driving knowledge specifically. Here it is hypothesised that individuals with higher levels of education will also hold more knowledge of eco-driving.

3.2.4   ECO-DRIVING SUPPORT

The eighth, and final, research question asks how much people are willing to pay for eco-driving support, be that in the form of training, or via in-vehicle support devices. Previous research suggests that willingness to pay for eco-driver training is generally low (King 2011), and even lower for in-vehicle devices. Respondents to Trommer and Hötl's (2012) survey study disagreed that these types of devices are worth paying for at all. It is therefore hypothesised that willingness to pay will be lower for in-vehicle support tools than for eco-driver training.

3.2.5   SUMMARY OF PURPOSE

As aforementioned, the ultimate aim of this chapter is to inform the design of suitable means for encouraging eco-driving, via the acquisition of information regarding the general public's knowledge and perceptions of the practice. Understanding what people already know of eco-driving and what they think of it, and how these variables relate to demographic factors and environmental attitudes, will help to provide focus and direction for any potential future investment, be that via training, educational initiatives, policy change, or technological development (e.g. in-vehicle devices). The following research questions are therefore posed (summarised from above):

(Q1) What perceptions do people have of eco-driving and its effects?
(Q2) What do people know of eco-driving (i.e. of the specific behaviours)?
(Q3) Are more pro-environmental individuals more knowledgeable of the means for eco-driving?

(Q4) Do more pro-environmental individuals report performing eco-driving behaviours to a greater extent than less pro-environmental individuals?

(Q5) Do people with greater knowledge of eco-driving also report performing it to a greater extent?

(Q6) How does knowledge of, and propensity to perform eco-driving behaviours vary with age and gender?

(Q7) Do those with higher levels of general education also have more knowledge of eco-driving behaviours?

(Q8) How much are people willing to pay for eco-driver training and in-vehicle, eco-driving support devices?

3.3 SURVEY

The questionnaire was administered online through the University of Southampton's iSurvey online questionnaire tool, and consisted of three main sections, in the following order; (1) demographics, (2) eco-driving awareness, knowledge, and perceptions, and (3) environmental attitudes. Demographics questions included age (18–24, 25–34, 35–44, 45–54, 55–64, 65 or over), gender (male or female), and level of highest completed education (GCSE or equivalent, A level or equivalent, undergraduate degree, postgraduate degree, none of the above), and asked whether or not individuals had received additional or advanced driver training (yes or no). This section also asked participants the number of years they had held a licence, the amount driven annually, and whether or not they currently had access to a vehicle (see Appendix C for the full survey).

3.3.1 Eco-driving Section

3.3.1.1 Perceptions

In the eco-driving section, participants were first told 'the way in which a car is driven affects the amount of fuel consumed per mile'. They were then asked to provide estimates of this effect for others ('about how much difference do you think this "driving behaviour" can have for the average person?') and for themselves ('what kind of effect do you think it could have for your fuel use?'). Eight possible responses were offered for each of the two questions; 0%–5%, 5%–10%, 10%–15%, 15%–20%, 20%–25%, 25%–30%, 30%–35%, more than 35%. Participants were then asked 'have you heard about the practice of "eco-driving"?' and 'what do you think of "eco-driving"?'; possible responses are displayed in Figures 3.3 and 3.4 in results Section 3.4.2.1.

3.3.1.2 Knowledge

In order to assess knowledge of the specific means for driving in a fuel-efficient manner (in a way that would be less prone to self-report biases than simply asking participants of their level of eco-driving knowledge), a line of questioning identical to that used by King (2011) was adopted. The section also assessed the participants' self-reported tendency to perform such behaviours. Participants were first asked 'Could you give a tip for reducing fuel consumption while driving?' (answers given in free text), followed by the question 'How often do you follow this advice?', the five responses to which ranged from 'never or almost never' to 'always or almost always'. This pair of questions was repeated four times (with the additional word 'another' for the second, third and fourth). The participant was told to skip each question if not able to provide an eco-driving tip.

3.3.2 ENVIRONMENTAL ATTITUDES

The third section assessed participants' environmental attitudes, and was taken directly from Harvey et al. (2013). Each of the 26 questions asked the participant to what extent they agreed with a given statement, with possible responses given on a 7-point Likert scale, ranging from 'strongly disagree' to 'strongly agree' (see Harvey et al. 2013, 6). All questions are displayed in Table 3.4, alongside results (Section 3.4.1.2).

3.3.3 PARTICIPANTS AND ETHICS

The sample was obtained in an opportunistic fashion; respondents were recruited through social media (e.g. Facebook, Twitter), email and paper flyers (indicating the web address of the survey) distributed around the Southampton area (in particular the University of Southampton). The snowball approach was used, relying on friends, family and acquaintances to pass on physical flyers, emails and weblinks. Ethical approval was sought from, and granted by the University of Southampton's Ethics and Research Governance committee, reference number 11243. Only those 18 years or over, who had held a full driving licence for at least 1 year, were asked to participate.

3.4 RESULTS

Three hundred and twenty-one people responded to the survey. Three hundred and eleven respondents are summarised, in terms of age and gender, in Table 3.1, alongside group percentages as proportions of the total sample size (i.e. 321). Ten participants did not provide age or gender information, corresponding to 3.1% of the sample. Only six respondents aged over 65 completed the survey (three male, three female); this group was therefore combined with the 55–64 group for all subsequent analyses, forming a '55 and over' group (reflected below in Table 3.1).

TABLE 3.1

Age and Gender Groups of the 311 Respondents to these Questions

		Age Group				
		18–24	25–34	35–44	45–54	55 and over
Gender	Female	18 (5.6%)	62 (19.3%)	19 (5.9%)	25 (7.8%)	36 (11.2%)
	Male	27 (8.4%)	67 (20.9%)	19 (5.9%)	19 (5.9%)	19 (5.9%)

The number of years with a driving licence ranged from 1 to 55 (M = 18.69, SD = 12.91), with 31 respondents stating that they did not currently have access to a vehicle. As may be expected from the sampling approach (i.e. university-centric) the average highest finished education level was higher in the sample than would be expected in the general population; 5 reported no qualifications, 25 at GCSE or equivalent, 54 at A level or equivalent, 124 undergraduate degree or equivalent and 112 at postgraduate degree or equivalent. One respondent did not answer this question. Four respondents stated that driving constituted the *main* part of their job, and 67 as a part of their job (but not the main part); the remainder did not drive for their work. The survey also attracted 5 fleet drivers (though 11 did not answer this yes or no question), 76 drivers with extra training (7 did not answer this question), and 16 drivers belonging to motoring organisations (4 did not answer this question).

3.4.1 Data Reduction

3.4.1.1 Eco-driving Knowledge

Of the 321 respondents, 60 did not provide any tips in the eco-driving knowledge section, corresponding to 18.7% of the sample. The remaining 261 respondents provided 723 distinct tips (provided in free text); these were coded according to the coding scheme presented in Table 3.2. The tips were split into two broad groups; 'valid' and 'invalid'. Valid tips were those that, if followed, could be reasonably assumed to increase efficiency. Invalid tips were those that, if followed, would not have a beneficial effect on efficiency. The coding scheme was predominantly developed from the responses themselves (i.e. iteratively); however, it was also informed by a general understanding of the behaviours that most influence fuel efficiency, developed through reviews of popular literature (from sites such as www.ecodrive. org), and the academic literature (Barkenbus 2010; Hooker 1988).

The questions in this section specifically asked about saving fuel 'while driving'; however, in the popular eco-driving literature (Ecowill 2015; Energy Saving Trust 2015; The AA 2015) strategies such as proper tyre inflation, route planning, weight and drag reduction and engine maintenance all commonly appear as eco-driving tips. Therefore, despite not being behaviours that can be performed *while driving*, they have been included here as valid eco-driving tips. Not only do they have significant effects on fuel economy, but they are also commonly reported as eco-driving behaviours in the types of easily-accessible media to which the majority of respondents could have been exposed.

The reader will note that speed choice was split into two categories in Table 3.2 – one valid, one invalid. This reflects the fact that simply driving slowly is not an eco-driving behaviour in and of itself; it is the avoidance of excessively high speeds, particularly on motorways (where high speeds are expected), or the maintenance of an optimum speed (Barth and Boriboonsomsin 2008), that represent eco-driving strategies. Unqualified statements referring to slow driving were therefore considered invalid. Statements suggesting compliance with speed limits were also coded under this invalid speed category. Not only is compliance with road laws expected (hence not considered an additional fuel-saving strategy), but also it does not usually

TABLE 3.2

Breakdown of Categories, with Examples from Responses

	Code	Example From Survey Responses
	Correct gear choice and rev minimisation	'Drive in higher gears'
	Gentle acceleration and braking – general	'Plan your breaking [sic], break [sic] gradually and accelerate gradually'
	Gentle acceleration and braking – acceleration specific	'Do not accelerate too fiercely'
	Gentle acceleration and braking – deceleration specific	'Reduce speed gradually instead of braking sharply'
	Avoidance of acceleration and braking – general	'Slowing the car without braking where possible and safe'
	Avoidance of acceleration and braking – use of momentum	'Use momentum and gravity e.g. coasting, declutching down hills etc.'
Valid	Avoidance of acceleration and braking – anticipation of traffic	'Look and plan ahead so you can drive progressively by timing your arrival with gaps in traffic'
	Avoidance of acceleration and braking – anticipation of road environment	'Cruise to junctions instead of hard braking'
	Drag reduction	'Don't drive with windows open'
	Avoid excessive speed	'Reduce average motorway speed to 60 mph'
	Minimise peripheral use	'Turn off the aircon'
	Avoiding idling	'Turn off the engine at long waits e.g. level crossings'
	Tyre inflation	'Keep tyres inflated'
	Weight reduction	'Take the rubbish out of your boot'
	Engine tuning and maintenance	'Ensure engine is well maintained'
	Route planning	'Take the car for multiple tasks/multiple people' in one journey'
	Slow driving	'Drive slowly'
Invalid	Avoid car use	'Walk when possible, or ride a bike'
	Other	'Don't listen to loud rock music'

save fuel (e.g. driving at 60 mph on a UK motorway is more efficient than driving at the 70 mph speed limit, and 45–50 mph is the most efficient speed regardless of the posted limit).

Advice related to the avoidance of car use altogether (e.g. 'cycle, walk, or use public transport where possible') was also coded as invalid. The questionnaire expressly discussed fuel efficiency and car use; although avoiding car use is arguably the most effective strategy for minimising fuel consumption, it is not an eco-driving tip.

Repeated advice was normally coded as invalid insofar as if a respondent provided more than one tip under a single code (e.g. two tips related to gear changes) only one of those was accepted. However, the two acceleration and deceleration codes ('gentle acceleration and braking' and 'avoidance of acceleration and braking') were each split, the first into two subdivisions, the second into three; this allowed recognition for responses specifying different aspects of the same general class of behaviour. The first author coded all provided tips.

In order to test the coding scheme, 72 (approximately 10%) of the distinct tips were chosen at random for inter-rater reliability testing. Tips coded as invalid due to repetition were replaced with another tip chosen at random. The final 72 tips were given to two other researchers, both personally known to the authors (both were based in the same Human Factors Research Unit at the University of Southampton). Neither was involved in the research project in any other way. Following a short training session on the coding scheme, each researcher independently coded the 72 segments (assigning one category to each). Percentage agreement between the first author's assignments and those of the first additional coder was 90.3%. Cohen's kappa was calculated, returning a value of 0.89 ($p < 0.001$), indicating a very good agreement (Altman 1991). The calculations made to compare the first author's own categorisations with those of the second additional coder revealed identical statistics; 90.3% agreement and a Cohen's kappa of 0.89 ($p < 0.001$). Each author/additional coder pairing saw seven disagreements; however, only one of these was common to both pairings. The tip 'change gear around 4,500 revs in a petrol vehicle', coded as incorrect by the first author (advice suggests changing gear between 1,500 and 2,500 rpm), was coded under the 'gear change and rev minimisation' category by both additional coders.

In an attempt to quantify eco-driving knowledge (in order to arrive at an eco-driving 'score' for each participant) each eco-driving tip was scored in terms of the potential effect of the described behaviour on fuel economy. The effect of a given behaviour on the efficiency of a vehicle will, of course, depend on a multitude of factors for example vehicle type, road topography and traffic conditions. Furthermore, estimates vary across the literature; for example, estimates for the effect of 'aggressive' driving on fuel use ranges from 25% (Sivak and Schoettle 2011) to 41% (De Vlieger 1997). For the purposes of obtaining a general eco-driving knowledge 'score', however, absolute values are less important than relative values. The purpose of the score was to assess people's eco-driving knowledge relative to other respondents, rather than relative to actual fuel-efficiency values (which are themselves highly dependent on context). Work by Sivak and Schoettle (2011) was therefore used to guide scoring, as these authors provide estimates for 9 of the 11 valid categories outlined in Table 3.2.

To have these estimates come from the same research effort, rather than across various articles, reduces potentially confounding effect of differences in measurement techniques or test vehicles. Sivak and Schoettle do not, however, provide estimates for the effect of optimal gear choice and engine revolution (rev) minimisation, nor for aerodynamic drag, and no single article could be found that provided estimates for all 11 categories. Values assigned to these two categories were therefore taken from other sources (see Table 3.3). This is recognised as a potential limitation of the scoring system.

For 'gear choice and rev minimisation', Beckx et al. (2007) provide an estimate of 30% for the difference in fuel consumption between 'normal' and 'aggressive'

TABLE 3.3

Coding Scheme, Assigned Scores and Sources from Which Scoring Was Developed

| | Potential Fuel Saving According to Source | | | |
Description	Sivak and Schoettle (2011)	Haworth and Symmonds (2001)	Beckx et al. (2007)	Assigned Score
Gear choice/engine revs			30%	30
Gentle brake/acceleration	25%[a]			25
Avoid brake/acceleration	25%[a]			25
Drag		10%–15%[b]		10
Speed	6%[c]			6
Route selection	6%[d]			6
Engine maintenance	4%			4
Peripherals	4%[e]			4
Idling	2%[f]			2
Tyre inflation	1.5%[g]			1.5
Weight reduction	1.5%[h]			1.5

[a] Saving refers to 'aggressive driving' compared to good practice.
[b] Savings made at high speed.
[c] Savings made if optimal speed choice on motorways for 20% of the route.
[d] Savings if choosing one of two route options available for 20% of the total distance driven.
[e] Savings made when avoiding use of air conditioning for 25% of distance driven.
[f] Savings achieved by turning off engine during two 1-minute idle periods every 10 miles.
[g] Savings made compared to under-inflation of all four tyres by 5 psi.
[h] Savings made with 100 pounds (45.4 kg) of excess weight removed.

gear change assumptions, averaged across an entire journey. For the effects of aerodynamic drag, Haworth and Symmonds (2001) provide an estimate of 10%–15% for vehicles travelling at high speeds. The lower end of this estimate has been used as the score for this category in order to reflect the potential savings made across an entire journey rather than for a section of a journey (in line with other categories). Regarding the 'gentle acceleration and braking' and 'avoidance of acceleration and braking' categories, participants were only assigned a score for each of the subcategories provided they had not already provided a 'general' comment for that category. Generous coding was, however, applied; if the respondent provided, for example, one general comment, and then one comment each for two different subcategories, they were scored for two responses, that is, they were given the maximum allowable.

All invalid tips were given a score of 0.

Finally, this section also asked about the respondents' propensity to follow their own eco-driving advice. In order to arrive at a single score for each participant the mean of the reported propensity values (1 = 'never or almost never', 5 = 'always or almost always') corresponding to each respondent's *valid* tips was calculated. Propensity scores for invalid tips were not included in the calculation.

3.4.1.2 Environmental Attitudes

As aforementioned, the environmental attitudes section was taken directly from Harvey et al. (2013), the questions and results for which can be found in Table 3.4. Each of the 26 items invited the respondent to indicate, on a 7-point Likert scale, the extent to which they agreed with a given statement (from 1, strongly disagree, to 7, strongly agree). Principal component analysis with varimax rotation was undertaken, resulting in the identification of four factors. Although Harvey et al. (2013)

TABLE 3.4
Postfactor Analysis Environmental Attitude Section Results

Factors	F1	F2	F3	F4	Item Mean	Item SD
F1: General Energy Use Attitudes						
Q2: Need to find better ways to produce clean energy	**0.665**	0.138	0.218	−0.075	6.41	0.88
Q4: We live in an energy-guzzling society	**0.573**	0.008	0.463	−0.118	6.04	1.03
Q5: Energy issues are overrated	**−0.614**	−0.067	0.050	0.226	2.51	1.50
Q7: High energy consumption is bad for the environment	**0.492**	−0.004	0.119	−0.102	5.98	1.14
Q8: I do not see how we can make large reductions in fuel and energy use	**−0.547**	−0.148	0.187	−0.160	2.91	1.56
Q14: Worry that gas and oil will run out in 30 years	**0.315**	0.192	0.158	0.259	4.46	1.67
Q21: I would support congestion charges to help reduce traffic and pollution	**0.432**	0.233	0.187	0.224	4.00	1.86
Q24: It's a waste of time to get people to use cars less	**−0.501**	0.028	0.162	−0.421	3.52	1.70
F2: Energy Conservation Attitudes						
Q1: Motivated to save money on energy at home	0.280	**0.586**	0.144	−0.040	6.05	0.94
Q11: When I next buy a car, I will choose one with better fuel consumption	0.189	**0.453**	−0.005	0.131	5.19	1.50
Q15: At home, like to get cheapest energy possible	−0.173	**0.675**	−0.002	−0.107	4.71	1.54
Q16: Try to reduce energy consumption at home	0.186	**0.744**	0.070	0.153	5.73	0.98

(Continued)

TABLE 3.4 (Continued)
Postfactor Analysis Environmental Attitude Section Results

Factors	F1	F2	F3	F4	Item Mean	Item SD
Q18: Switch off lights wherever not used	0.067	**0.575**	0.055	0.074	5.93	1.03
Q22: Wasting energy annoys me	0.380	**0.495**	0.179	0.231	5.63	1.10
F3: Incentives For Energy Use Reductions						
Q3: People at work don't care about saving fuel	−0.104	0.051	**0.684**	−0.118	4.96	1.55
Q12: Energy prices must rise to sort out problems	0.095	0.029	**0.439**	0.073	3.97	1.75
Q13: People will only change energy consuming habits forced to	0.190	−0.034	**0.596**	−0.003	5.17	1.44
Q17: People will only save fuel if they have an incentive	−0.187	0.219	**0.680**	0.123	5.07	1.37
Q20: Traffic fumes in city centres bother me	0.332	0.383	**0.319**	0.034	5.09	1.51
Q26: People care more about saving fuel at home than at work	−0.072	0.207	**0.583**	0.300	5.90	1.05
F4: Motivation To Use Public Transport						
Q9: I would travel on public transport if it were cheaper	−0.053	0.075	0.026	**0.775**	4.76	2.01
Q23: I would travel on public transport if it were more convenient	0.071	−0.080	0.153	**0.831**	5.68	1.57
α Coefficients	0.666	0.681	0.626	0.695		
Removed Items						
Q6: My own contribution to saving fuel could be better	0.310	−0.346	0.111	0.127	5.35	1.21
Q19: It's important to complete a journey as quickly as possible	−0.244	−0.154	0.145	−0.067	3.37	1.59
Q25: I would only buy an eco car if it were no more expensive than a normal car	−0.394	0.061	0.116	0.118	4.78	1.70
Q10: I like to keep check on my car's mpg	−0.064	0.412	0.090	−0.069	4.76	1.92

also reported four factors, the analysis presented here resulted in different item groupings: F1, general energy use attitudes; F2, energy conservation attitudes; F3, incentives for energy use reductions; and F4, motivation to use public transport.

Factor 1 initially grouped items 2, 4, 6, 7, 8, 19, 21, 24 and 25, resulting in a Cronbach's alpha coefficient (Cronbach 1951) of 0.655; however, after removal of items 6, 19 and 25 (due to low item-total correlations) the alpha for this scale rose to 0.666. Similarly, factor 2 initially comprised items 1, 10, 11, 15, 16, 18 and 22, resulting in an alpha of 0.653; removal of item 10 resulted in an alpha of 0.681. Factor 3's low alpha score of 0.626 could not be remedied by removal of any items. Factor 4 achieved an alpha value of 0.695. These reliability scores are not high (Nunnally 1978); however, the scale used here is a short one. Not only are high alpha values harder to achieve with shorter scales (Streiner 2003), the 0.7 threshold itself (particularly when used for shorter scale lengths) has been questioned (Schmitt 1996). These factors were therefore accepted; however, it is important to note that short scale length does not alleviate problems of unreliability, therefore all following analyses made with factor scores are done so with caution.

3.4.2 ADDRESSING RESEARCH QUESTIONS

3.4.2.1 Q1: What Perceptions Do People Have of Eco-driving and Its Effects?

Respondents were first provided with the statement 'the way in which a car is driven affects the amount of fuel consumed per mile'. They were then asked about the size of this effect, for both themselves (personal effect), and for others (effect for population). One individual (of the 321 respondents) did not answer this question. Results are presented in Figure 3.1; here it can be seen that the estimated differences were higher for 'the average person' (effect for population) than for the respondents' own fuel use (personal effect). This suggests that respondents consider that a change in driving style would result in larger differences in other peoples' fuel-efficiency measures than in their own. A Wilcoxon signed-ranks test was applied; this revealed the difference between responses for 'personal effect' and 'effect for population' to be statistically significant ($Z = -6.74$, $p < 0.001$).

The term 'eco-driving' was then introduced, and the respondents asked about their knowledge and perceptions of the practice. Figures 3.2 and 3.3 display percentages of responses to the two items regarding this question. All 321 respondents answered both of these questions (i.e. no missing data). Only one response was allowed for the question 'Have you heard about the practice of "eco-driving"?' (Figure 3.2). For the question 'What do you think of "eco-driving"?' (Figure 3.3), the respondent was invited to tick all that apply (hence in Figure 3.3 proportions combine to over 100%).

From Figure 3.2 it can be seen that of those who have heard of eco-driving, the majority state that they 'have an idea of how to do it' (114 individuals, 35.5% of the total sample); however, 77 individuals (24.0%) stated that they had not heard of it. From Figure 3.3 it is clear that the respondents generally consider eco-driving positively, with three lower options in Figure 3.3 receiving by far the greatest number

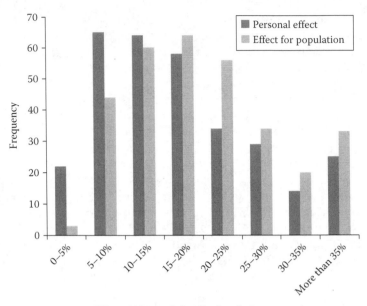

Effect of 'driving behaviour' on fuel consumption

FIGURE 3.1 Frequencies of responses to statements regarding the effect of driving behaviour on fuel consumption.

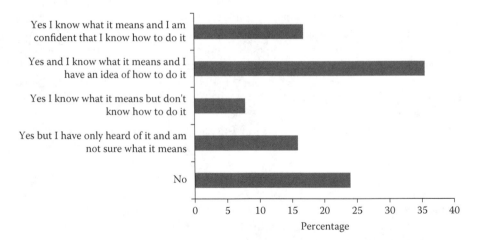

FIGURE 3.2 Percentage of responses to the question 'Have you heard about the practice of "eco-driving"?'

of responses. Only two respondents considered the practice to be unsafe, and only seven selected 'The UK/the world does not need it'. However, 28 respondents indicated that eco-driving reduces driving enjoyment too much, and 34 indicated that time pressure is more important. Here, 75 individuals (23.4%) stated that they had not heard of it.

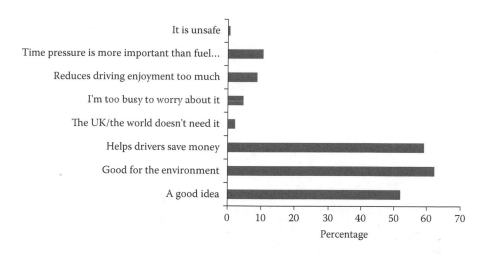

FIGURE 3.3 Percentage of respondents selecting answers to the question 'What do you think of "eco-driving"? (Please tick all that apply)'.

3.4.2.2 Q2: What Do People Know of Eco-driving (i.e. of the Specific Behaviours)?

As aforementioned, 261 of the 321 respondents (81.3%) provided at least one eco-driving tip (including both valid and invalid tips), while 60 (18.7%) did not provide any tips at all (i.e. they left this section blank). A total of 46 individuals provided one tip (14.3%), 63 provided two (19.6%), 57 provided three (17.8%) and 95 (29.6%) provided four, the maximum possible. In total, 723 distinct tips were provided; 629 (87.0%) of these were coded as valid, 94 (13.0%) as invalid. 58 respondents (16.8%) provided the maximum number of valid tips (i.e. four), while 13 respondents (4.0%) provided only invalid tips. Most of those offering at least one tip, however, provided no invalid tips at all (180 respondents, 56.1%). Figure 3.4 displays the number of respondents providing each possible number of tips, separated into valid tips and invalid tips.

Regarding the *types* of tips most commonly provided, Figure 3.5 displays the frequencies with which each of the tip categories appeared. It can be seen that 'gear choice and rev minimisation' was the most commonly reported single category, with 112 individuals (34.9%) offering tips under this category. When combining subcategories (which are presented separately in Figure 3.5), we see 122 instances of the 'gentle acceleration and deceleration' category, and 154 instances of the 'avoidance of acceleration and deceleration' category. These do not, however, correspond to 122 and 154 respondents, as each respondent was able to provide a tip in each of the subcategories without incurring a 'repetition' categorisation (see Section 3.2.1). Only 109 respondents (34.0%) provided one or more tips under the 'gentle acceleration and deceleration' category, and 131 respondents (40.8%) provided one or more tips under the 'avoid acceleration and deceleration category'. This final category therefore represents the type of eco-driving advice reported by the greatest number of respondents. Of the invalid tips, 'slow driving' was referred to most often; 38 respondents (11.8%) gave tips under this category.

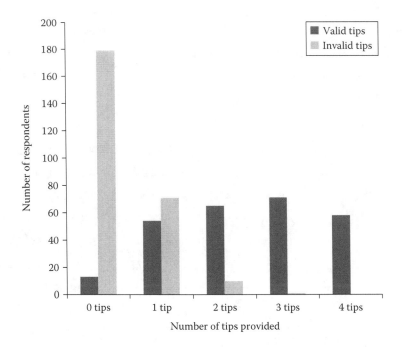

FIGURE 3.4 Number of respondents providing the different possible numbers of valid and invalid tips.

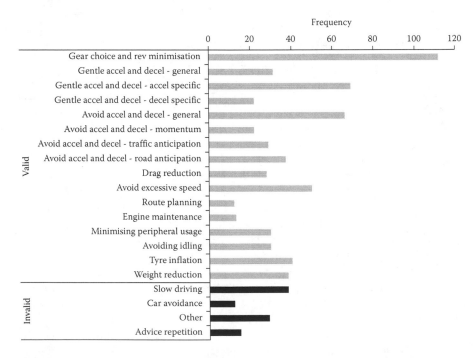

FIGURE 3.5 Frequency of tips, by category.

With regard to tip score, values ranged from 0 (for the 13 respondents providing only invalid tips, and the 60 respondents providing no tips at all) to 105, the maximum possible score (i.e. one gear choice comment, and three comments under the 'gentle acceleration and deceleration' or 'avoidance of acceleration and deceleration' categories). Three respondents (0.9%) attained this maximum score. The mean tip score was 35.08 (SD = 27.84).

To assess whether or not self-reported knowledge of eco-driving actually corresponds to measured knowledge, tip score (measured knowledge) was assessed in respect of respondents' answers to the question 'Have you heard about the practice of "eco-driving"?' (self-reported knowledge, with possible answers including 'No', 'Yes, but only heard of it', 'Yes, but don't know how to do it', 'Yes, and have an idea of how to do it' and 'Yes, and confident how to do it'; see Figure 3.2, above). Of the 321 respondents for whom tip score could be calculated, 3 did not answer the self-reported knowledge question, hence sample size for these calculations was 318. The Kruskal–Wallis test revealed significant differences in tip score between the five groups ($H = 32.12$, $p < 0.001$). Jonkheere's test revealed a significant trend in the data; tip score increased across answers from 'No' to 'Yes, and confident how to do it', $J = 23387$, $z = 4.74$, $p < 0.001$. Differences between group means for those providing responses to first three options were, however, small. Those indicating that they have an idea how to perform the practice, or those confident of how to do it, scored considerably higher. Group means and standard deviations are presented in Table 3.5.

3.4.2.3 Q3: Are More Pro-environmental Individuals More Knowledgeable of the Means for Eco-driving?

Firstly, it is useful to note that agreement with the reality of the current global sustainability challenge was relatively high among the respondents; the majority of respondents agreed with the need to 'find better ways to produce clean energy' (M = 6.41, SD = 0.88, on a scale of 1 to 7), and generally disagreed with the statement 'energy issues are overrated' (M = 2.51, SD = 1.50). Furthermore, agreement with the statement 'I am motivated to save money on energy consumption at home' was also high (M = 6.05, SD = 0.94).

To assess relationships between environmental attitudes and eco-driving knowledge, normalised scores for factors 1 and 2 ('general energy use attitudes' and 'energy

---

TABLE 3.5

Means (and Standard Deviations) for Tip Score, Grouped by Responses to the Statements 'Have You Heard of Eco-driving?'

	No	Yes, but only heard of it	Yes, but don't know how	Yes, and have an idea how	Yes, and am confident how
Mean	27.54	24.13	24.77	43.77	44.33
SD	23.68	24.28	29.13	27.51	28.33

conservation attitudes'), for each respondent, were calculated (i.e. the average across the items in each scale). The relationships these variables had with tip score was then assessed using Spearman's ρ. This statistic is computed based on ranks, rather than true values, and its use is recommended when data do not meet conditions of normality (Field 2009). Due to various combinations of missing data, the sample sizes for each of the three calculations (and those in subsequent sections) were slightly different (i.e. only full data sets were included in each calculation); sample size for each calculation is therefore presented. First, factor 1 and factor 2 were moderately correlated ($\rho = 0.37$, $p < 0.001$, $N = 288$). Factor 1 was significantly correlated with tip score; however, effect size was not high ($\rho = 0.24$, $p < 0.001$, $N = 294$). Similarly, factor 2 was also significantly correlated with tip score, though again only weakly ($\rho = 0.21$, $p < 0.001$, $N = 307$).

3.4.2.4   Q4: Do More Pro-environmental Individuals Report Performing Eco-driving Behaviours to a Greater Extent than Less Pro-environmental Individuals?

As aforementioned, to calculate a score indicating the general propensity to perform eco-driving behaviours (i.e. to follow one's own advice) the average of the respondents' individual propensity scores across all provided tips (from 0 = low propensity, to 4 = high propensity) was calculated. As described above, only propensity scores relating to valid tips were included in the calculation. The sample size therefore reduced to 248; propensity scores could not be calculated for the 13 respondents providing only invalid tips, nor for the 60 providing no tips at all. Mean propensity score was 2.99 (SD = 0.78), with values ranging from 0 to 4.

Scores for factors 1 and 2 were again used to investigate the relationships between environmental attitudes and the propensity to perform eco-driving behaviours. Using Spearman's ρ, it was found that factor 1 scores did not significantly correlate with propensity scores ($\rho = 0.07$, $p = 0.310$, $N = 232$). A significant correlation was, however, found between factor 2 scores and propensity score; this relationship was a moderate one ($\rho = 0.24$, $p < 0.001$, $N = 237$).

3.4.2.5   Q5: Do People with Greater Knowledge of Eco-driving Also Report Performing it to a Greater Extent?

To address this question Spearman's ρ was calculated for the relationship between propensity score and tip score; this revealed a significant but weak relationship ($\rho = 0.14$, $p = 0.035$, $N = 248$).

3.4.2.6   Q6: How Does Knowledge of and Propensity to Perform Eco-driving Behaviours Vary with Age and Gender?

To address this research question, two 2 (gender) × 5 (age; see Table 3.2 above) factorial ANOVAs were performed; one for tip score, the other for propensity score. Both sets of data met conditions of normality (visually checked using histograms) and equality of variance (Levene's test resulted in significance values of 0.943 and 0.365 for tip score and propensity respectively). For tip score, a main effect for gender was found; $F_{(1,301)} = 18.30$, $p < 0.001$, partial $\eta^2 = 0.057$. Subsequent pairwise comparison revealed that males (M = 43.19, SD = 30.61) scored significantly higher than females (M = 28.44, SD = 30.14; $p < 0.001$). No main effect for age was found ($F_{(4,301)} = 0.75$, $p = 0.844$, partial $\eta^2 = 0.0035$), nor was there a significant interaction effect ($F_{(4,301)} = 0.75$, $p = 0.561$, partial $\eta^2 = 0.010$).

Similar results were found for propensity scores; the 2 × 5 ANOVA revealed a significant main effect for gender ($F_{(1,229)} = 11.46$, $p = 0.001$, partial $\eta^2 = 0.048$) but no significant main effect for age ($F_{(4,229)} = 2.37$, $p = 0.053$, partial $\eta^2 = 0.040$). For propensity, however, a significant interaction effect *was* found; $F_{(4,229)} = 3.12$, $p = 0.016$, partial $\eta^2 = 0.052$. In terms of gender differences, a pairwise comparison revealed that males provided significantly higher propensity scores (M = 3.18, SD = 0.73) than females (M = 2.81, SD = 0.81). With regard to the interaction effect, this is represented in Figure 3.6. Greater differences between age groups for females than for males were observed; however, the non-linear trend in scores for females across age groups is not one that invites a simple explanation.

3.4.2.7 Q7: Do Those with Higher Levels of General Education Also Have More Knowledge of Eco-driving Behaviours?

As data once again satisfied conditions of normality (visually assessed) and equality of variance (Levene's test resulting in a significance value of 0.180), a one-way ANOVA for tip score was performed to assess the relationship between tip score and

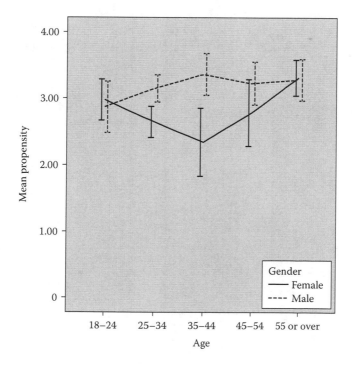

FIGURE 3.6 Interaction effect between age and gender for self-reported propensity to follow one's own eco-driving advice, with 95% confidence intervals indicated.

general education across five groups (GCSE or equivalent, A level or equivalent, undergraduate degree, postgraduate degree, none of the above). This revealed no significant effects; $F_{(4,256)} = 2.38$, $p = 0.052$, partial $\eta^2 = 0.029$.

3.4.2.8 Q8: How Much Are People Willing to Pay for Eco-driver Training and In-vehicle, Eco-driving Support Devices?

Respondents were asked to indicate how much they would be willing to pay for eco-driver training, and how much they would be willing to pay for an in-vehicle eco-driving support tool, each on a 6-point scale from 'Nothing' to 'Over £300'. The driver training question received 302 responses (94.1%) and the in-vehicle device question received 312 (97.2%) responses. Results are displayed in Figure 3.7. A Wilcoxon signed-rank test revealed a significant difference between the results of the two questions ($Z = -5.75$, $p < 0.001$). The median response of willingness to pay for an in-vehicle device was 'Up to £50'; however, for professional training the median was 'Nothing'. For both questions the 'Nothing' category received the most responses; for in-vehicle devices, 121 respondents (38.8% of those who responded) stated that they would pay nothing, while for driver training 168 respondents (55.6%) stated that they would pay nothing.

3.5 DISCUSSION

This survey study has attempted to address a variety of research questions, some of which were exploratory in nature, other of which invited specific hypotheses. Results will therefore be discussed in relation to each research question in turn.

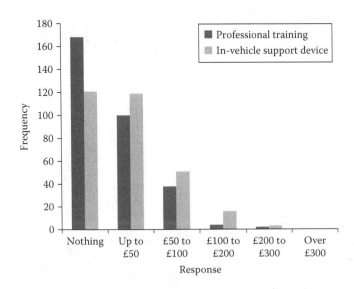

FIGURE 3.7 Frequency of responses to the questions regarding willingness to pay for both professional eco-driving training and for in-vehicle eco-driving support devices.

3.5.1   Q1: What Perceptions Do People Have of Eco-driving and Its Effects?

Results suggested that, on the whole, people hold positive views of eco-driving, mirroring results from King's survey of New Zealand AA members (King 2011). This positive result implies that people may be willing to adopt the technique given the right incentives. Many respondents indicated that the practice is not only good for the environment, but also that it helps drivers to save money. Hence any eco-driving encouragement incentives would do well to encapsulate both global environmental *and* personal financial benefits, rather than focus on one alone.

In King (2011) results indicated that people were more interested in learning defensive driving (for safety) than learning eco-driving techniques; however, the two have significant overlaps (e.g. Young et al. 2011), and very few respondents to this chapter's survey (only two) considered eco-driving as unsafe. We conclude, therefore, that safety concerns are not a barrier to eco-driving uptake; however, given the importance of safety in driving (a highly safety-critical domain), training a combination of the two approaches is likely to be beneficial.

Regarding the potential effect of eco-driving on efficiency that is, the savings that can realistically be achieved by altering driving behaviours, respondents' estimates were largely in accordance with the 10% posited in the literature (Barkenbus 2010), at least for the effect that the practice would have on their *own* fuel consumption. In terms of the effect that the practice would have on *other* drivers' efficiencies, respondents estimated greater savings than they would themselves achieve. This could be indicative of the oft-cited overconfidence bias (Alicke 1985; Alicke et al. 1995); people consider themselves to already be more efficient than the average driver, and therefore have less room for improvement (as was suggested by results in King 2011). An alternative explanation, however, is that people think that other drivers may be more successful at eco-driving, and thereby achieve greater savings. Further study would be required to untangle this issue; however, regardless of the underlying reason for this result, the difference in estimates is important for any eco-driving education or advertisement programme, as it implies that care should be taken to stress the effectiveness for *all* drivers, regardless of how they currently perceive themselves. For example, when given accurate information about the performance of others, overconfidence effects are attenuated (e.g. Moore and Small 2007).

3.5.2   Q2: What Do People Know of Eco-driving (i.e. of the Specific Behaviours)?

In King (2011) the most commonly reported eco-driving tip was in relation to light acceleration and braking. Results from this chapter repeat this finding, although more specifically regarding the *avoidance* of acceleration and deceleration (a further specification not made in King's research). Furthermore, 'gear choice and rev minimisation' also featured highly. These three most commonly reported categories reflect the relative importance of each behaviour in terms of their effects on fuel consumption as reported in the literature (Barkenbus 2010; Hooker 1988; Sivak and

Schoettle 2011), suggesting that those able to provide eco-driving tips are generally aware of the behaviours that most influence fuel efficiency.

Additionally, these categories also generally reflect those found by Franke et al. (2016) in their interview study of highly efficient Toyota Prius drivers. Although their study focused on the challenges brought specifically by hybrid vehicles, many of the eco-driving strategies were the same as those offered by respondents to this chapter's research. However, whereas Franke et al.'s participants could be considered experts in fuel-efficient driving (i.e. they were recruited from an online fuel monitoring website, and each recorded above-average efficiency values), respondents to this chapter's survey were not. It is perhaps unsurprising, therefore, that the absolute numbers of respondents providing valid eco-driving advice was quite low. Fewer than half of the respondents mentioned 'gear choice or rev minimisation' as a behaviour that influences fuel economy, with similar findings for both 'gentle acceleration and deceleration' and 'avoidance of acceleration and deceleration'. This suggests that education and training will have an important role to play in encouraging eco-driving; one cannot perform a pro-environmental behaviour without first knowing the possible action strategies that are available (e.g. Hines et al. 1987; Jensen 2002).

The results presented above also indicate that people are aware of their own levels of knowledge, at least in a relative sense (in general accordance with Lichtenstein and Fischhoff 1977). That is to say, the trend in tip score across respondents was also seen across self-reported knowledge of eco-driving; those who think they are more knowledgeable do indeed score more highly (on tip score) than those who think themselves less knowledgeable. For training and education purposes this is important, as when judging where incentives should be focused it may be sufficient to simply ask people how much they know of eco-driving, rather than assessing knowledge via some form of test. This result does not, however, tell us about *absolute* knowledge, and as the group means in Table 3.5 show, even those in the two higher scoring groups (i.e. those how answered that they had an idea, or were confident, of how to eco-drive) still achieved a tip score (on average) considerably lower than the maximum possible. Hence, although some may be more deserving of eco-driver training or education than others, such a scheme would likely benefit all participants.

3.5.3   Q3: ARE MORE PRO-ENVIRONMENTAL INDIVIDUALS MORE KNOWLEDGEABLE OF THE MEANS FOR ECO-DRIVING?

In order to measure the relationship between eco-driving knowledge and pro-environmental attitudes, only factors 1 and 2 of the environmental attitude section were used ('general energy use attitudes' and 'energy conservation attitudes'). Factor 3 was not employed, as not only was the reliability of this scale quite low (with a Cronbach's alpha of 0.626), but also it related more to opinions on incentives for energy reduction rather than on general energy use and conservation attitudes. Factor 4 was not used as this only dealt with motivations to use public transport. Moreover, it is worth pointing out again that the four factors identified here were different to those identified by Harvey et al. (2013) in their use of the same questionnaire. It is therefore more difficult to make comparisons between the results of this chapter and those of Harvey et al. (ibid.). That general energy use attitudes and energy conservation attitudes are significantly related was, however, a general finding common to both this chapter's results and those of Harvey et al. (ibid.). In terms of the relationship these two factors have with eco-driving knowledge, although the two correlations were significant, effect sizes were weak. This reflects Arcury's (1990) finding of a weak relationship between environmental knowledge and attitudes (Flamm 2006). We therefore tentatively accept the hypothesis that those with more pro-environmental attitudes are also more knowledgeable of the means for eco-driving; however, with the caveats of a weak effect size, and the potential problems associated with relatively low-reliability scales (i.e. factor 1 with an alpha of 0.666, factor 2 with an alpha of 0.681).

3.5.4 Q4: Do More Pro-environmental Individuals Report Performing Eco-driving Behaviours to a Greater Extent Than Less Pro-environmental Individuals?

As above, only factors 1 and 2 were used to address this research question, this time looking at propensity score. Previous literature suggests that pro-environmental individuals are more likely to engage in pro-environmental behaviours (Bamberg and Möser 2007), and that more pro-environmental individuals report a higher frequency of performing eco-driving behaviours (Delhomme et al. 2013); regarding this relationship, results from this survey were mixed. First, factor 1 scores, those relating to general energy use attitudes, were not at all related to self-reported propensity to eco-drive, in contrast to Delhomme et al.'s (2013) findings. Factor 2 and propensity scores, on the other hand, did show a significant, moderate relationship; those scoring higher in energy conservation attitudes also reported a higher propensity to perform eco-driving behaviours. This is in contrast to Harvey et al.'s (2013) finding that energy attitudes and eco-driving are not conceptually linked; this chapter's results suggest that those who report performing energy conservation strategies to a greater extent (with all but one of the items referring to domestic or general conservation rather than travel choices) are also more likely to report they perform eco-driving. Given the mixed results, however, we cannot accept the hypothesis that those with more pro-environmental attitudes will also report a higher propensity to perform eco-driving behaviours.

3.5.5 Q5: Do People with Greater Knowledge of Eco-driving Also Report Performing It to a Greater Extent?

As aforementioned, Hines et al. (1987; see also Jensen 2002) suggested that a self-reported performance of pro-environmental behaviours is associated with a greater knowledge of available action strategies. The significant correlation between tip score and propensity score shown above lends tentative support to this assertion, though given the small effect size (Spearman's ρ of just 0.14), strong conclusions are difficult to make. We can, therefore, only cautiously accept the hypothesis that those who are more knowledgeable of eco-driving behaviours also report a higher propensity to perform them.

3.5.6   Q6: How Do Knowledge of and Propensity to Perform Eco-driving Behaviours Vary with Age and Gender?

As described in the introduction, the relevant literature regarding this question is mixed; King (2011) found that older males were more knowledgeable about eco-driving, whereas Delhomme et al. (2013) found that older *females* were more likely to *report* performing such behaviours. In terms of the age question, the results presented above suggest there are no differences across age groups, hence do not support either of these previous findings, nor allow us to accept the hypothesis that older drivers are more knowledgeable of, and more likely to report they perform eco-driving behaviours.

Regarding gender, it was found that males were both more knowledgeable of eco-driving behaviours, and were more likely to report performing them. This may be a reflection of the well-established and pervasive gender differences in car culture (O'Connell 1998). Although males were, traditionally, more intensive car users, the situation is changing; with more gender equality in the labour force comes intensified car use across both genders (Best and Lanzendorf 2005). Any scheme or programme intending to encourage eco-driving should therefore ensure that it reaches female drivers as well as male drivers.

3.5.7   Q7: Do Those with Higher Levels of General Education Also Have More Knowledge of Eco-driving Behaviours?

This question aimed to assess whether or not Diamantopoulos et al.'s (2003) finding, that people with higher levels of education were also more knowledgeable about environmental issues, could also be extended to knowledge of eco-driving strategies. The results presented above suggest that it cannot; no significant differences in tip score between groups of differing levels of education were found. We therefore reject the hypothesis that more highly educated individuals will also be more knowledgeable of eco-driving.

3.5.8   Q8: How Much Are People Willing to Pay for Eco-driver Training and In-vehicle, Eco-driving Support Devices?

The final research question posed in the introduction asked about people's willingness to pay for either eco-driver training, or for an in-vehicle eco-driving support device. According to King (2011), people are not willing to pay a realistic sum for eco-driver training, and following Trommer and Hötl (2012), people are willing to pay nothing at all for in-vehicle devices. This led to the hypothesis that respondents would be willing to invest more in the former than the latter. In fact, the above results suggest that the opposite is true; on average, people are willing to pay more for in-vehicle device than they are for additional driver training. Importantly, however, and in accordance with both studies cited above, respondents reported low willingness to pay for either.

With regard to training, this suggests that including eco-driving in pre-licence training may be the most effective way to teach the technique, rather than relying on post-licence training uptake. Indeed, to this end the Driver and Vehicle Standards Agency (DVSA; a UK governmental agency) includes in its official *DVSA Guide to Driving* (an industry-standard text recommended especially to learner drivers) a full chapter on 'ecosafe' driving, the practice of safely minimising fuel consumption in the vehicle (Driver and Vehicle Standards Agency 2015). However, according to a report from the Driving Standards Agency (the pre-2014 name for the DVSA), criticism has been voiced regarding the way in which eco-driving is included in the standard driving test, particularly with regard to consistency (Campbell-Hall and Dalziel 2011). This report also suggested that post-licence training specifically focusing on eco-driving was not only rarely offered, but was more difficult to sell than training offering eco-driving tips *alongside* other advanced driving content. A more suitable method to train efficient drivers may therefore be to stress the overlaps between safe and efficient driving (of which there are many; e.g. Young et al. 2011). Such an approach may also satisfy respondents to King's (2011) survey; here, the training of defensive driving was seen as more desirable than the training of eco-driving techniques (though see also Delhomme et al. (2010) for an argument for using environmental protection concerns to encourage safer driving).

Finally, for in-vehicle support tools, we would argue that it is primarily up to the car manufacturers to invest in implementation, rather than relying on consumers to invest in their own, additional devices. Although people are generally not willing to pay extra for such devices, there is a considerable body of literature suggesting that they can, and do, help drivers to save fuel (e.g. Birrell et al. 2013, 2014a; Jamson et al. 2013; Muñoz-Organero and Magaña 2013; Staubach et al. 2014b; Van der Voort et al. 2001; Wada et al. 2011). Moreover, it is difficult to imagine how the more complex eco-driving support systems, such as those that integrate information from various vehicle sensors, mapping data and engine performance logs in order to support anticipatory driving (Continental 2015), could possibly be implemented *without* car manufacturing companies taking the lead.

3.5.9 STUDY LIMITATIONS

The first major limitation to this research is the fact that it relies on self-reports of behaviour and attitudes. In terms of environmental behaviours, some have argued that self-reports are adequate indicators of actual behaviour (Fuj et al. 1985; Warriner et al. 1984), whereas others have expressed doubts (Corral-Verdudo 1997). When looking at driving behaviour specifically, Lajunen and Summala (2003) found that social biases (which were expected to influence self-reports) did *not* significantly affect results. Importantly, however, they did not assess actual, observed behaviour. We therefore accept this as a limitation of the study design.

Although we have attempted to place an objective value on eco-driving knowledge, any attempt to quantify such knowledge, and represent it in a single figure, is prone to inaccuracies and biases. High inter-rater reliability suggested an adequate coding scheme; however, the effect of some eco-driving behaviours is highly dependent on context, and on the intensity with which they are performed. For example, although the avoidance of harsh acceleration is universally recommended as an eco-driving practice, to accelerate *too* slowly can incur greater fuel consumption than positive, firm acceleration. Also, it may be the case that some respondents, when referring to 'slow driving' (an invalid tip), were actually aware that this is only a valid fuel-saving technique at high speeds, yet did not explicitly indicate this when responding (e.g. they may have assumed that this would be understood). The categorisation scheme presented above could not have captured these potential differences, and although we have attempted to be as objective as possible in the assessment of eco-driving knowledge (thereby arriving at a score that truly does reflect the knowledge of the individual), we accept the inability to *definitively* do so as a limitation of the study. We also accept as a limitation the upper limit of four eco-driving tips. This number was chosen as it replicated King's (2011) approach; however, it is quite possible that some participants would have provided more than four tips if given the opportunity.

Finally, it is important to bring to attention the limitations of the sample. First, 73.8% of the sample held an undergraduate degree-level qualification or higher. This is far higher than the 24% found in the UK's adult population as a whole (Office for National Statistics 2012). Second, the over 65 group was under-represented in the sample (at under 2%), likely a result of the combination of the survey's web-based design, and the snowball approach to sampling. Additionally, all of the respondents indicated living in the UK; we cannot, therefore, generalise to other nationalities or cultures. As Harvey et al. (2013) demonstrated, attitudes towards eco-driving, and indeed towards the transport system as a whole, differ with nationality. How this chapter's results might differ across culture is a question that can be addressed only by future research.

3.5.10   GENERAL DISCUSSION

In terms of the knowledge of eco-driving held by the public at large, there is general correspondence between the behaviours most commonly reported as having an effect on fuel economy, and those argued (in the literature) to be most influential (i.e. gear choice, and acceleration and deceleration behaviours). Furthermore, people's estimates of the potential savings that eco-driving practices can bring about are similar to those seen in the academic literature, with general consensus that eco-driving is a worthwhile practice that helps drivers to save fuel, and benefits the environment. However, overall knowledge of specific eco-driving behaviours was quite low, and the relationships between environmental attitudes and knowledge of, and propensity to perform eco-driving behaviours, although they exist, are weak. Neither pro-environmental attitudes nor a good knowledge of eco-driving behaviours was strongly linked with a propensity to perform eco-driving. Indeed, the determinants of pro-environmental behaviours are highly complex, and there exists no single framework that can fully explain the relationships between attitudes, knowledge and behaviours (Kollmuss and Agyeman 2002); we would argue that this is also true for eco-driving specifically. Encouraging uptake of the practice will, therefore, require a mixed approach. Interventions aimed solely at one aspect, be it environmental attitudes, general environmental knowledge or eco-driving specifically, will not suffice.

Although results from this study do not suggest that the teaching of eco-driving behaviours will itself encourage their uptake, we still argue that this would be worthwhile [following Hines et al. (1987) and Jensen (2002)]. This should, however, be done alongside a continued education programme about the environment as a whole. We agree with Barkenbus (2010), that driver eco-training should be subsidised, and argue that it should be included in pre-test training; it can be effective, but people are not willing to pay for it, as evidenced in our results. The same is true for in-vehicle, eco-driving support devices; although shown in the literature to be effective, people are simply not willing to pay extra for them. We argue, therefore, that it is largely up to car manufacturers to integrate such devices into new vehicles.

Although eco-driving on its own may appear to have a relatively minor impact on society's total energy use (when we also consider industrial, commercial and domestic domains), it is a mitigation strategy requiring almost no change to infrastructure or technology hence is relatively inexpensive to implement, and yet can still realise around 10% savings for the average driver (Barkenbus 2010). To achieve widespread uptake of eco-driving will, however, require a concerted effort on the part of government and industry, as well as the end-user.

3.6   CONCLUSIONS

The purpose of the research presented in this chapter was to inform the potential means for teaching and encouraging eco-driving in the general population; however, it has not focused on any one particular strategy, (e.g. pre-test training or in-vehicle information devices). Rather, it has addressed more generally the public's knowledge of, and attitudes towards eco-driving as a practice, how these interact with environmental attitudes more widely, and how these variables can guide interventions aimed at promoting fuel-efficient driving. The fact that attitudes towards eco-driving are generally positive is promising; however, knowledge of specific behavioural strategies was shown to be quite low, with only around half of the respondents able to provide eco-driving tips related to the three most influential classes of behaviour.

These results suggest that there is merit in guiding drivers in their performance of the specific behaviours characteristic of eco-driving, behaviours that the drivers themselves would not exhibit spontaneously *without* guidance. This chapter has, however, used only survey data; it can tell us little of the cognitive strategies actually used while in control of the vehicle, and does not reveal the potential differences in cognitive strategies employed by more or less efficient drivers. If we are to design suitable in-vehicle information that supports such behaviours (the ultimate aim of the research presented in this book) it is useful not only to understand the general public's knowledge of, and attitudes towards efficient driving, but also the specific cognitive strategies employed by drivers when driving 'naturally'. In other words, it would be useful to investigate what people are actually thinking when they drive. To understand such links would help to us discover where problems lie (in less economical drivers) and where drivers perform well (in more economical drivers), knowledge that would help guide in-vehicle information system design. This is the focus of the next chapter.

4 Verbal Reports
An Exploratory On-road Study

4.1 INTRODUCTION

As has been discussed previously in this book, it has long been known that a person's driving style has a significant effect on the fuel economy of the vehicle (e.g. Evans 1979; Waters and Laker 1980). Indeed, estimates of the difference between the consumption rates of different driving styles run as high as 30% (Bingham et al. 2012). The academic community is now largely aware of the source of these differences, at least in terms of the individual behaviours that characterise fuel-efficient driving (in particular, gear change timings, and acceleration and deceleration strategies; e.g. Barkenbus 2010; Hooker 1988). Moreover, it seems that the general public are generally aware of these behaviours, and report a willingness to perform them (Chapter 3). What is less clear is the relationship between individual drivers' cognitive strategies, their overall awareness and their fuel economy.

In terms of the overall journey of this book, the current chapter presents a further narrowing of attention. Chapter 2 provided an initial investigation of the broad topics of energy use, the effect of design on behaviour, and of the potential for ergonomics and human factors to have an impact in the uptake and energy-efficient use of low-carbon vehicles. As was discussed, focus moved away from the low-carbon vehicle specifically, towards the efficient use of *any* private road vehicle. In particular, Chapter 3 introduced and investigated eco-driving as a term that encompasses the practices that characterise efficient use of the vehicle. In Chapter 3 a broad perspective of eco-driving was taken, with attention paid to the general public's perceptions, attitudes and knowledge of the practice. The current chapter, taking a more directed approach, represents a step towards one of the ultimate aims of the research presented in this book; to design and test an in-vehicle, eco-driving support system. To move towards this goal, it is first useful to develop an understanding of what cognitive processes, or strategies, are used by more or less efficient drivers. This is the aim of the current chapter; to arrive at a better understanding of what it is that make some drivers more efficient than others. Such an understanding, it is hoped, will go on to inform the design of in-vehicle information that supports those *less* adept at driving efficiently.

4.2 VERBAL PROTOCOL ANALYSIS

It is, of course, impossible to directly observe the cognitive processes of an individual performing a task. Recording one's eye movements, or observing one's physical behaviours, gestures, and interactions with task artefacts both offer data sets that can be interpreted in terms of underlying cognition (e.g. Rayner 1997; Underwood et al. 2002). However, to collect these forms of data while an individual is engaged in the driving task can be costly (in time and resources), distract from the task at hand (and therefore change cognition), and provide data sets that are difficult to interpret (in terms of cognitive mechanisms and structures). One such method that attempts to circumvent these issues is verbal protocol analysis (Ericsson and Simon 1980, 1993).

The technique has two primary variants; concurrent think-aloud and retrospective think-aloud. Concurrent think-aloud requires an individual to verbalise their thoughts concurrently with task performance; in retrospective think-aloud the participant provides verbal reports after the task has finished. Each has its advantages and disadvantages, and each may be more suited to certain environments or domains (e.g. Banks et al. 2014b; Russo et al. 1989; Van den Haak et al. 2003).

Ericsson and Simon (1980) speak of different types, or levels, of verbalisations that may be produced. They do so in terms of the potential for intervening, recoding processes to occur in the time between information leaving the central processor (in working memory; e.g. Atkinson and Shiffrin 1968; Baddeley and Hitch 1974) and the production of the verbalisation. Three levels of verbalisation are specifically described. Level one represents a direct reproduction of information from working memory; this undergoes no intermediate processing. Level two involves the recoding of non-verbal internal representations into verbal code. This does require an additional degree of translation but crucially does not alter a person's cognitive processes (Ericsson and Simon 1993); the individual is not expected to explain their thoughts, nor is attention directed or manipulated by the researcher. The third level does just this; it requires either the explanation (rather than simply reporting) of thoughts, a scanning or filtering of thoughts (for particular referents in the environment or task), or a necessity to verbalise aspects to which the participant would not normally attend. Ericsson and Simon explain this in terms of the difference between explaining and thinking aloud: 'When subjects are asked for explanations, the reports cannot be generated without extending the information and relations heeded [...] Hence, thinking aloud, as distinguished from explanation, will not change the structure and course of the task processes' (1980, p. 226). It is, therefore, verbalisations at levels one and two in which we are interested when seeking to understand basic cognitive processes and structures. Hence the approach used in this study was to use concurrent think-aloud without directing the participants' attention to any particular objects or events, requiring the participants to simply *report* their thoughts as opposed to *explaining* them.

As aforementioned both concurrent and retrospective reporting have advantages and disadvantages, and both may be suited to particular tasks or environments (that is, Russo et al. 1989). It has been argued that retrospective reporting suffers from problems of non-veridicality, that is, the lack of correspondence between verbal reports and cognitive processes, particularly in tasks of long durations (e.g. Van Gog et al. 2009). The driving sessions in the study presented in this chapter lasted for just under 20 minutes (comparable to the average journey time of 23.7 minutes in the UK; Department for Transport 2014). Retrospective verbal reports of tasks of this length are not only subject to issues of omissions and post-task fabrications (Van Gog et al. 2009), but can be biased towards positive aspects (Swann et al. 1987) and effective actions (Kuusela and Paul 2000). Moreover, Nisbett and Wilson (1977) demonstrated that participants providing retrospective verbalisations are no more accurate in the identification of the determinants of behaviours than are observers. We therefore argue the concurrent think-aloud procedure to be a more appropriate technique in the context of the current study.

There are a number of examples in the literature of the concurrent think-aloud procedure as applied to the driving domain. One early example comes from Hughes and Cole (1986) in their investigation of the foci of attention when driving. In this study, however, one could argue that the verbal protocol analysis procedure (as argued for by Ericsson and Simon 1980, 1993) was not strictly followed, insofar as participants were specifically asked to report that which drew their attention (hence resulting in the potential for level three verbalisations). Indeed, when talking of the intermediate scanning or filtering of thought processes, Ericsson and Simon (1980) themselves talk of the driving domain, making reference to Soliday and Allen's work (see Ericsson and Simon 1980, 219) in which participants were asked to report all perceived traffic hazards. The act of directing attention requires the participant to first scan the environment and then categorise that which they see before verbalising their thoughts, thereby representing level three verbalisations. Therefore, these cannot be said to accurately reflect underlying cognitive processes.

More recently, Walker et al. (2001a) reported an on-road investigation in which concurrent verbal reports contributed to an understanding of the role of feedback in driver cognition, and Lansdown (2002) used concurrent verbal reporting in a driving simulator study to investigate individual differences in drivers, highlighting differences between novices and experts. It is possible to find additional such studies more recently still; for example, Walker and colleagues furnish us with two more uses of the concurrent think-aloud procedure in the driving domain. In Walker et al. (2011) the technique was again applied in an on-road setting, with the aim of investigating the differences in the ways car drivers and motorcyclists interpret road situations. Results highlighted some 'critical incompatibilities' (Walker et al. 2011, 878) between the two road user groups. In Walker et al. (2013) the technique was used for the analysis of situation awareness, also in an on-road environment; however, the participants here were asked to 'explain their actions' (Walker et al. 2013, 21), hence may have suffered from the presence of level three verbalisations (following Ericsson and Simon's arguments).

Further examples of the on-road use of concurrent verbal reporting can be found in work by Young and colleagues (2013, 2015). In both studies, the first of which investigated distraction-induced driver error (Young et al. 2013), the second investigating attention at railway crossings (Young et al. 2015), verbal protocol analysis was used as an additional source of information, enriching the data set built up from interviews, videos and the output of the instrumented vehicle. Banks et al. (2014b) also used the concurrent think-aloud procedure as an additional source of information in a driving simulator study. In their study, an exploratory analysis of verbal reports was used in conjunction with quantitative simulator data, with the verbal reports adding value to the understanding of the thought processes that underlie behavioural outcomes. Finally, Pampel et al. (2015) provide, to our knowledge, the only instance of a driving study to use the concurrent think-aloud procedure for the specific investigation of eco-driving. The authors compared the verbal reports of people when asked to drive 'normally', 'safely', and 'efficiently', drawing on mental model research in discussing the simulator-based study's results. The study's findings led the authors to argue for the existence of mental models specific to eco-driving; models that are not employed when asked to drive 'normally'. The authors drew attention to the existence of misconceptions concerning speed and travel time. Whereas Pampel et al. (2015) were interested in the differences exhibited by participants when driving for different guiding purposes (i.e. normal, safe, or efficient), this chapter is interested simply in the potential of verbal reports to help build an understanding of what distinguishes more efficient drivers from less efficient drivers, without any guidance of behaviour or driving style.

Specifically, it deals with the possible differences between those who are more or less fuel-efficient when driving 'normally', that is, driving without additional instruction or guidance. The question is, therefore; do drivers who display more fuel-efficient driving styles (as evidenced by quantitative vehicle data) also show differences in their underlying cognitive structures and processes (as revealed in their concurrent verbal reports)? This question is, of course, highly exploratory in nature, and does not invite specific hypotheses; we simply seek to identify common strategies held by those who exhibit more efficient driving styles, and investigate the ways in which they might differ from those who drive in a less economical manner.

4.3   METHOD

The data used in this chapter come from research undertaken at the University of the Sunshine Coast, Australia, the aim of which was to assess the effect of providing concurrent verbal reports on driving performance (see Thomas et al. 2015). Although subsequent analyses are entirely distinct, the study procedure and materials described below match those reported in Thomas et al. (2015).

4.3.1   Participants

Participants were recruited through newsletter lists maintained by the University of the Sunshine Coast research team, and were compensated with a $50 (Australian) voucher for their time. Ethics approval was sought from and granted by the University of the Sunshine Coast Human Ethics Committee. Although 20 participants completed the study (see Thomas et al. 2015), data for 1 participant (number 18) were not used in the current analyses. Not only were weather conditions significantly different for this participant than for any other (they experienced the only instance of rain), they were not a native English speaker (hence carry the possibility for an extra level of translation from thought to speech). Of the 19 participants whose data were analysed, 9 were male, 10 were female, with ages ranging from 28 to 49 (M = 39.00, SD = 6.10), years with licence from 11 to 32 (M = 21.42, SD = 6.28), and weekly kilometres driven from 70 to 750 (M = 366.84, SD = 215.87).

4.3.2 APPARATUS

All participants drove the same standard Ford Focus 2.0L Trend sedan test vehicle, with automatic transmission. The Centre for Human Factors and Sociotechnical Systems' on-road capability (ORCa) is instrumented with Racelogic's video VBOX pro, a system that collects and stores data from two high-definition cameras, a micro phone, GPS, an accelerometer and the vehicle's CAN bus. Data are captured at a rate of 10 Hz and includes speed, steering angle, brake and accelerator pedal inputs, handbrake position and engine speed.

4.3.3 PROCEDURE

Participants were required to drive a 15 km urban route, located in the suburbs around the University of the Sunshine Coast, six times; twice for familiarisation, twice silently, twice while providing verbal reports. The same route was driven each time, and all driving sessions took place between 10 am and 2 pm on weekdays.

The participants were first trained to provide concurrent verbal reports; they were provided with verbal instructions on the technique, following which they spent 10 minutes in a driving simulator practising the act of verbalising their thoughts. An experimenter provided guidance and feedback (on the act of verbalisation) throughout this practice session. Participants were then introduced to the route to be driven, initially via a paper map. Two familiarisation laps (in which the participant was accompanied by an experimenter) were then performed. In the first, the experimenter provided verbal route guidance. In the second, guidance was provided only upon request.

Then followed the experimental trials, in which the participant drove unaccompanied. Ten participants provided verbal reports in the first two laps, driving in silence in the third and fourth laps; this order was reversed for the remaining 10 participants. This chapter uses data from only the first of the two laps involving verbalisations; therefore, when we include the familiarisation session, the data analysed here come from the third driving session for 9 participants, and from the fifth driving session for 10 participants.

4.3.4 DATA REDUCTION

4.3.4.1 Vehicle Data

Due to the lack of a direct measure of fuel consumption in the available ORCa data, it was necessary to calculate statistics that could provide an indirect indication of driving efficiency. Although all participants drove at approximately the same time of day (as aforementioned), it was of course impossible to completely control for traffic condition variability. Hence it was not possible to use the standard deviation of vehicle speed as such a proxy for efficiency (as argued by Birrell et al. 2014); this would be too significantly affected by changing traffic conditions between participants. Using vehicle speed (in kilometres per hour) and throttle input (in percentage depression, 0%–100%) two statistics were calculated that we would argue suggest a fuel-efficient driving style; time spent coasting and excessive acceleration.

The first statistic, time spent coasting, is indicative of a participant's tendency to take advantage of the vehicle's momentum when driving. Although the term 'coasting' is used here, this does not imply that the participant put the vehicle into neutral, rather it is simply the act of travelling forward (i.e. at speeds greater than 0) with zero accelerator pedal depression (see, e.g. Staubach et al. 2014a; Hajek et al. 2011) for fuel savings brought about by systems supporting such behaviour). The raw statistic was not used, as this would also be too heavily influenced by traffic conditions and general driving speed. For some participants the route took longer to drive, hence those having driven for longer would have more time in which to exhibit coasting behaviours. The measure has, therefore, been expressed as a percentage, that is, the time spent travelling forward without depressing the throttle as a percentage of the total time spent travelling forward.

The second statistic, excessive acceleration, comes from research reported in Birrell et al. (2013). In their study of vibrotactile eco-driving support, Birrell and colleagues argued that, for eco-driving, throttle use should not exceed a 50% depression threshold [arguments based on Johansson et al. (1999), and van der Burgwal and Gense (2002); both cited in Birrell et al. (2013)]. To measure excessive acceleration, Birrell et al. (2013) calculated the product of the magnitude of throttle position when depressed beyond 50% and the time spent over that threshold. In other words, the measure represents the area above the 50% depression line but under the curve of a graph made by plotting throttle depression percentage by time. This measure is, however, still affected by traffic variability; more traffic would necessitate more stop and start behaviours, hence more acceleration events. This has also, therefore, been expressed as a percentage, that is, the area of the curve above 50% and below the throttle depression/time curve as a percentage of the total area under the curve. We would argue that this provides a measure of a participant's tendency to depress the throttle excessively when performing manoeuvres requiring acceleration, and should therefore be unaffected by the gross number of acceleration events.

4.3.4.2   Verbal Data

Each of the 19 videos (recorded using the VBOX system described above) used in the analyses were initially transcribed verbatim, the transcripts of which were then segmented into single identifiable units of meaningful speech. As is to be expected, some participants uttered more words than others, with total word counts ranging from 1262 to 3532 (M = 2119.63, SD = 724.06). The average length of a unit of speech across a single participant's transcript also varied between participants, ranging from 6.54 to 12.61 words (M = 9.21, SD = 1.48). The transcripts were initially subjected to a coding scheme based on the information processing functions described in Banks et al. (2014a), namely monitor, anticipate, detect, recognise, decide, select and respond, thereby representing a theory-driven, top-down approach. Subsequent development and refinement of the coding scheme then proceeded in a purely bottom-up, data-driven fashion. The coding scheme was developed using the first two transcripts, after which it was applied to the following two transcripts. The scheme was altered such that all four transcripts were adequately categorised; it then was applied to the subsequent two transcripts. This iterative process continued until all transcripts were adequately categorised.

Only one individual (i.e. the first author of this book) categorised all transcripts; no other researchers were involved in the coding process. The issue of intercoder reliability is therefore of less significance than if, for example, some of the transcripts were coded by one researcher and others by a different researcher. It was important, however, to measure the extent to which the codes created actually reflected the text, or, in other words, whether or not they made sense. One of the transcripts was therefore given to two other researchers, both personally known to the current authors (both being based at the University of Southampton's Transportation Research Group). Both were experienced human factors and ergonomics researchers; however, neither was involved in the research presented in this chapter in any other way. The coding scheme was introduced to each researcher, and a description of each code was provided to them. A training period of approximately half an hour was required to fully explain the coding scheme, with examples from the transcripts.

One transcript in its entirety (that of participant seven) was used for the exercise. This transcript contained 1616 words, broken down into 185 statements. Average word count was 2119.63 words, and average number of statements was 230.32, hence participant seven's transcript represents one that was shorter than average, but was not the shortest of the sample (this was participant 20's, at 1378 words, categorised into 126 statements).

The level of agreement between the first additional coder and the principle investigator was calculated using the percentage agreement method. This resulted in a figure of 86.1%. Cohen's kappa was also calculated, and returned a value of 0.85, $p < 0.0005$, indicating a very good level of agreement (Altman 1991). The same calculations were performed to compare results between the second additional coder and the principle investigator, resulting in 92% agreement and a Cohen's kappa of 0.91 ($p < 0.0005$), again indicating very good agreement.

The resulting scheme consisted of 39 distinct codes, organised into six groups. The sixth group covered only non-driving related comments and inaudible or incomplete (i.e. unclassifiable) comments. The remaining five groups belonged to one of two higher-level categories, *action* or *state*. Table 4.1 displays the full coding scheme. The first group in Table 4.1 encompasses actions that were not obviously in preparation of something. These included simple statements of current action (without qualifying remarks; e.g. 'on to the roundabout' – participant 10) and statements describing actions performed in response to some change in the environment (other road users, general traffic or surroundings; e.g. 'just gonna increase a bit to get up this hill' – participant 3).

In contrast, the second group in Table 4.1 describes actions that were clearly in preparation of an upcoming event or change in environment (or indeed perceived event or change). This group includes statements concerning actions performed in order to reduce a subsequent real, perceived, or potential consequence, and statements describing actions made in order to reduce, or remove entirely, the need to act further down the road. We have included headway maintenance actions here (under code 2.6 in Table 4.1) as we consider the purpose of this behaviour to be inherently preparatory; one maintains a safe headway in order to give time and space to react safely, or indeed efficiently, to future environmental change (e.g. should the lead vehicle brake suddenly). A degree of subjectivity was impossible to avoid in assigning some statements to either the first or second code group as it was not always clear whether or not

TABLE 4.1

Coding Scheme with Examples from Transcripts

Category	Code	Subcode		Example
Actions	1. General actions	1.1. Avoid hazard or obstacle		'gotta move over a bit more because there's a trailer parked a bit close to the line' Ppt 11
		1.2. Give way or provide space to other road user	1.2.1. Decelerate	'just decelerating for the car in front' Ppt 2
			1.2.2. Accelerate	'I'm bringing my speed up so that I can give that pink car behind me more space' Ppt 13
			1.2.3. Wait	'waiting for the Volvo' Ppt 5
			1.2.4. Lane position	'move over to the right just out of courtesy to let them merge safely' Ppt 6
		1.3. Manage speed in response to surroundings	1.3.1. Decelerate	'slowing down, adjusting the speed for the road works' Ppt 6
			1.3.2. Accelerate	'and resuming back to the speed limit at the end of the road works' Ppt 6
			1.3.3. Act to maintain speed	'just making sure I get up this hill, it's quite steep' Ppt 3
			1.3.4. Generally maintain speed	'I'm just gonna drive really slowly' Ppt 3
		1.4. Directional		'turning right at the roundabout' Ppt 1
		1.5. Wait		'just now waiting for the red to turn to green obviously' Ppt 3
		1.6. Use momentum		'so I'm gonna stop accelerating now and let the car roll into the road' Ppt 15
		1.7. Indicate behaviour		'indicating left to leave the roundabout' Ppt 18
		1.8. Manage road position		'I'm gonna go into the far right lane' Ppt 7
	2. Preparatory or anticipatory actions	2.1. Accelerate		'so I'm just gonna speed up now because I know I have to merge into this, um, motorway traffic' Ppt 20
		2.2. Decelerate		'so I'm just decelerating to come round the corner' Ppt 2
		2.3. Manage road position		'staying in that left lane so I don't get caught in the right lane again and have to come over' Ppt 9
		2.4. Prepare for potential or real need to act		'just covering my brake just to make sure that no one jumps out' Ppt 9

(Continued)

TABLE 4.1 (Continued)
Coding Scheme with Examples from Transcripts

Category	Code	Subcode		Example
		2.5. Act to reduce potential consequences		'lots of cars parked ahead of me so just still going 20' Ppt 16
		2.6. Anticipatory response to behaviour of others		'so I am braking in order to keep my distance behind him' Ppt 19
	3. Actively monitor, check or search	3.1. General search		'looking for anyone that would be coming out' Ppt 1
		3.2. Monitor scene ahead		'looking over the hill as far as I can here' Ppt 5
		3.3. Monitor scene behind		'checking my rear vision mirror for the motorcyclist' Ppt 16
		3.4. Check own vehicle speed		'so just check that, um, my speed is right' Ppt 17
		3.5. Monitor specific road user		'just keeping an eye on the pedestrian on the left' Ppt 2
State	4. Description of current situation	4.1. Physical, fixed road environment		'there's a few streets coming off this road' Ppt 16
		4.2. Transient, temporary road environment		'heavy traffic coming the other way' Ppt 4
		4.3. Presence and behaviour of other vehicles		'a car on the left hand side trying to merge into traffic' Ppt 19
		4.4. Presence and behaviour of vulnerable road users		'mower guy on the right hand side doing mowing' Ppt 8
		4.5. Space and/ or time comment	4.5.1. General time/space to act	'time for me to move into this lane' Ppt 2
			4.5.2. Time/space to lead vehicle	'still got a good distance to the car in front of me' Ppt 15
			4.5.3. Time/ space to vehicle behind	'there's a good distance between the Mitsubishi behind me' Ppt 14
		4.6. Route and event knowledge		'these lights always take too long' Ppt 10
		4.7. Own vehicle status (including speed)		'and I'm close to the speed limit right now' Ppt 13
	5. Description of anticipated situation	5.1. Potential hazard appraisal		'there could be uni students walking through' Ppt 12
		5.2. Upcoming event		'just approaching the roundabout' Ppt 3
		5.3. Anticipate behaviour of others		'so the traffic that is lined up is gonna start moving' Ppt 5
Other	6. Other	6.1. Non-driving related		'Do I get a lolly? No, just a survey. Ok' Ppt 4
		6.2. Inaudible or incomplete		N/A

an action was performed as a necessary response to environmental change or whether it was an action aimed at a reducing subsequent need to act. For example, slowing down to negotiate a road curvature could either be seen as a necessary response to a road environment change, or as a preparatory action performed in order to reduce the subsequent requirement for harsh braking (upon entering the corner). To code such statements reference was made to the video data, with the distance from the event at which the comment was made playing a guiding role in code assignment. We accept this as a limitation of both the coding system, and of the methodology as a whole.

The third group, also under the *action* category, describes checking, monitoring or searching behaviours. These all denote *active* searches for information from the environment. General comments, for example, 'keep an eye out' (participant 1), were included under this group, as were mentions of specifically focusing on a specific road user or a search for space into which to act. This group also included speed checks (3.4), mirror and shoulder checks ('monitor scene behind'; 3.3), and monitoring of the scene ahead. The 'monitor scene ahead' (3.2) code was only applied when the participant *specifically* mentioned looking ahead or down the road. If a code was ambiguous it was preferentially assigned the code 'general search' (3.1). Many of the comments under this code involved a search for space into which to act, and relatedly, a search for the presence or absence of other road users.

The aforementioned groups are each concerned with *actions*, that is to say the participant is describing the performance of a particular behaviour; the second broad category deals with comments regarding the *state* of the environment. These are more passive in nature, and represent descriptions of that which the driver can see, and that which the driver expects, or recognises the possibility of seeing. This includes statements such as those describing the physical road environment, the presence of signage or road markings, the general state of the traffic situation, the presence and behaviour of other road users, and the space currently around the participant's vehicle. This also includes statements describing the vehicle's speed (as opposed to mentioning the *act* of checking the speed) and other vehicle systems. Furthermore, we have included in this category statements based on the participants' local knowledge (e.g. it being a school day or not) or experience of the general area, and of the route itself (e.g. that it is easy to speed on a particular hill). These comments, though conceptually different to the other subcodes in this group, are still descriptions of the current environment rather than anticipated events.

Comments regarding anticipated events comprise the second code under the *state* category, 'description of anticipated situations' (code 5). The first subcode, 'potential hazard appraisal' (5.1), applies to descriptions of states that may or may not exist, but for which the participant recognises the possibility, for example knowing that stationary vehicles in a car park may, at some point, start to move. The code does not include actions of preparation or avoidance, rather simply the description of the potential situation.

The second subcode, 'upcoming event' (5.2), encompasses statements concerning an upcoming change in the road environment that will (or may) necessitate action. To come under this code, rather than 'physical road environment' (4.1) for example, the statement had to include specific reference to the fact that the driver had not yet reached it, and would do so in the (near) future. Although one might argue that to *approach* something, or to have an event *coming up*, are statements more akin to actions than states (at least linguistically), we have included them under *state* as they refer to the existence of something in the upcoming road environment. The action of *approaching*, we would argue, is of less interest than is the object being approached, and the fact that the driver has observed, and noted, something to which they will have to react. Furthermore, this subcode includes statements such as 'ahead there are a set of traffic lights' (participant 5). These are more clearly *states* rather than *actions*, but refer to the same concept, that is, a future event in the road environment.

Finally, the code 'anticipate behaviour of others' (5.3) covers statements that refer to the participant's expectations about, or assumptions of the future or possible behaviour of other road users. These are *states* insofar as they are not *actions*; though the state to which they refer does not yet exist (or may never exist, being that this code also covers *potential* behaviour of others) it is a state nonetheless.

4.4 RESULTS

4.4.1 VERBAL PROTOCOLS

The frequencies with which each subcode appears in the 19 transcripts analysed are presented in Figure 4.1. Inset into the main graph is a pie chart indicating the

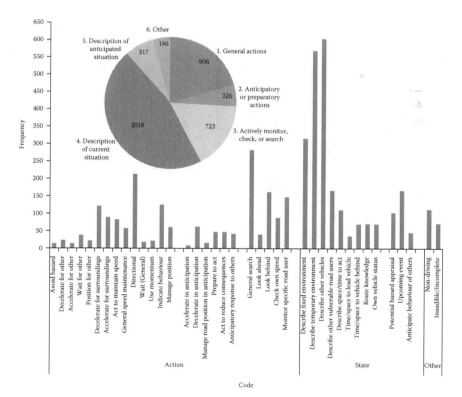

FIGURE 4.1 Frequencies of code assignations across all 19 transcripts, by subcode (horizontal axis), and by code (inset).

frequency with which each main code was applied. In total, 1855 statements were assigned a code under the *action* category, and 2335 assigned a code under the *state* category, indicating a bias towards descriptions of the current state of the environment, that is, describing what can be seen (or expected), over references to the behaviours currently being performed. Indeed, the vast majority of these statements fall under the 'description of current situation' code, closely followed by 'transient, temporary road environment', with the subcodes 'transient, temporary road environment' and 'presence and behaviour of other vehicles' receiving the most assignations.

Of the subcodes falling under the 'action' category, the two most commonly applied were 'general search' (under the 'actively monitor' group) and 'directional' (under the 'general actions' group). The 'preparatory or anticipatory actions' group was the least frequently applied of the five main groups (excluding the 'other' group). When use of this group was merited, it was the 'deceleration' subcode that was most commonly applied.

From Figure 4.1, five subcodes stand out as being particularly common. These five most commonly applied subcodes, namely 'directional' (1.4), 'general search' (3.1), 'describe fixed environment' (4.1), 'describe temporary environment' (4.2), and 'describe other vehicles' (4.3), accounted for 45.41% of all statements produced by the participants.

4.4.2 Vehicle Data

Time to complete the driving session ranged from 944.60 to 1175.00 seconds ($M = 1091.24$, $SD = 52.01$), with average speeds ranging from 44.62 to 55.08 kilometres per hour ($M = 48.17$, $SD = 2.40$). Instances of excessive acceleration were generally very low, with four participants never exceeding the 50% depression threshold. Average throttle depression ranged from 6.12% to 8.77% ($M = 7.68$, $SD = 0.615$), and maximum input ranged from 43.00% to 100% ($M = 61.29$, $SD = 17.72$). The average proportion of excessive acceleration was only 0.508% ($SD = 0.757$), with values ranging from 0% to 2.52%. In contrast, all participants spent a significant proportion of the time travelling with zero throttle depression, with values ranging from 36.65% to 51.16% ($M = 42.48$, $SD = 4.07$). Although these two measures, that is, coasting and excessive acceleration, have above been argued to both be indicative of a more fuel-efficient driving style, there was no significant correlation between them (Pearson's $r = 0.194$, $p = 0.425$).

4.4.3 Group Differences

In order to assess potential differences in the verbal reports made by different groups of drivers the participants were first split into distinct groups. The first split was based on experience, that is, those participants for whom it was their third driving session, and those for whom it was their fifth (lower experience, $n = 9$, and higher experience, $n = 10$). The second split, a median split, was based on the proportion spent coasting, resulting in low ($n = 9$, $M = 39.5\%$, $SD = 1.62$) and high ($n = 10$, $M = 45.2\%$, $SD = 3.66$) coasting groups. The third, again a median split, was based on the excessive acceleration variable, once again giving two groups; low (n = 9, M = 0.021%, SD = 0.027) and high (n = 10, M = 1.05%, SD = 0.813) excessive acceleration. As aforementioned, excessive acceleration and proportion coasting did not correlate; however, there were five participants that belonged to both the low excessive acceleration group and to the high coasting group. These 5 participants were therefore grouped, and compared to the remaining 14 participants. Table 4.2 displays ages, years with licence and distance driven each week for each of the groups; Figures 4.2 through 4.5 display the average code proportions (in percentages) for each of the groups.

As can be seen from the graphs, there are few discernible differences between the patterns of verbalisations of the different groups. A difference might perhaps be noticed between the proportion of statements assigned to the 'monitor, search' and 'describe situation' codes in the low and high coasting groups, and in the proportions of the 'describe situation' code between the low and high excessive acceleration groups; however, little else is immediately clear. In order to statistically assess any potential differences, a number of Mann–Whitney U tests were performed, the results of which are presented in Table 4.3. As can be seen from the table, no differences were found at all for any of the codes, between any of the groups, even without applying corrections for multiple comparisons (note that the 'other' code was not included in this analysis).

Upon closer inspection of the 'upcoming event' subcode (5.2), one that might be expected to feature highly in transcripts of the high coasting group, a small difference in the expected direction was indeed revealed. This difference was, however, very small; on average 3.87% of statements were attributed to this code for the low coasting group, and for the high coasting group this average was 4.05% (Mann–Whitney U = 44.5, p = 0.967). This lack of a pattern was also found for subcode 1.6,

---

TABLE 4.2

Mean (and Standard Deviation) Ages, Years with Licence and Distance Driven Weekly, by Group Membership

Variable	Group	Age	Years with Licence	Distance Driven Weekly (km)	Gender Split
Excessive Acceleration	Low (n = 10)	39.60 (5.89)	21.60 (6.40)	296.00 (188.46)	3 males, 7 females
	High (n = 9)	38.33 (6.96)	21.22 (6.89)	445.56 (239.74)	6 males, 3 females
Coasting	Low (n = 9)	38.78 (4.89)	20.78 (5.38)	368.89 (217.28)	4 males, 5 females
	High (n = 10)	39.20 (7.57)	22.00 (7.52)	365.00 (237.50)	5 males, 5 females
High coast + low excess acceleration	Both (n = 5)	42.00 (5.92)	24.60 (6.02)	212.00 (173.98)	2 males, 3 females
	Neither (n = 14)	37.93 (6.24)	20.29 (6.41)	422.14 (215.13)	7 males, 7 females

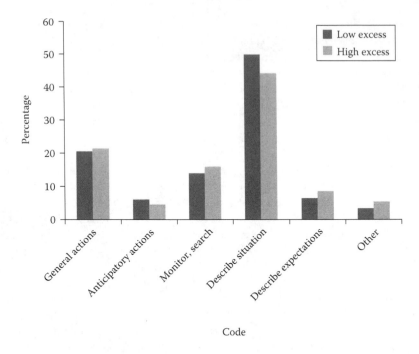

FIGURE 4.2 Code proportion by membership to low or high excessive acceleration groups.

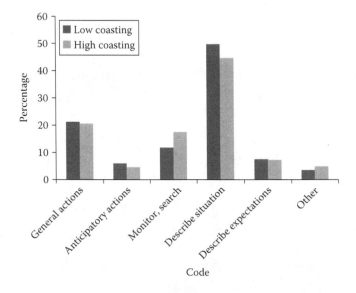

FIGURE 4.3 Code proportion by membership to low or high coasting groups.

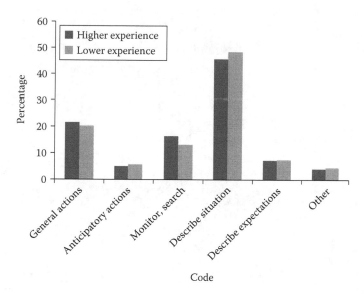

FIGURE 4.4 Code proportion by membership to low or high experience groups.

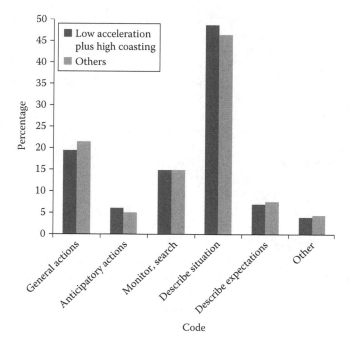

FIGURE 4.5 Code proportion by membership to both low excessive acceleration *and* high coasting groups, compared to the remaining participants.

TABLE 4.3

Mann–Whitney U Test Statistics (and Associated *p* Values) for Comparisons between Groups of the Proportions of Statements Assigned to Each Code Group

	General Actions	Anticipatory Actions	Monitor, Search	Describe Situation	Describe Expectations
Experience	39.0 (0.624)	38.0 (0.567)	43.0 (0.870)	39.0 (0.624)	37.0 (0.514)
Coasting	44.0 (0.935)	34.0 (0.513)	34.0 (0.369)	30.0 (0.221)	43.0 (0.870)
Excessive acceleration	43.0 (0.870)	32.5 (0.307)	44.0 (0.935)	37.0 (0.514)	28.0 (0.165)
Low accel + high coasting	32.0 (0.781)	31.5 (0.746)	31.0 (0.711)	35.0 (1.00)	34.0 (0.926)

'use momentum'. Again, it might be expected that those who perform more coasting behaviours also report doing so; this was not borne out in the results. Average proportions of statements assigned this subcode for the high (0.534%, SD = 0.969) and low (0.353%, SD = 0.589) coasting groups were not significantly different (Mann–Whitney $U = 44.0$, $p = 0.921$).

One might also expect those with more experience of the route (i.e. those for whom the session analysed was their fifth driving session, compared to those for whom it was their third) to produce more statements concerning upcoming events (given that they have more experience with the route). Again, this was not the case, and in fact the non-significant trend was in the opposite direction to that which might be expected, with the low experience group offering a slightly higher average proportion of these statements (4.16%) compared to the high experience group (3.79%).

In terms of excessive acceleration, rather than look at differences revealed by the coding scheme groups (or, rather, the lack of differences: see Table 4.2), results for the three subcodes specifically concerning acceleration were collated (i.e. codes 1.2.2, 1.3.2 and 2.1 in Table 4.1) and assessed. Although the high excessive acceleration group did produce a slightly higher proportion of statements referring to acceleration (M = 2.99, SD = 1.82) compared to the lower excessive acceleration group (M = 2.43, SD = 1.23), the difference was far from significant (Mann–Whitney $U = 42.5$, $p = 0.838$). Moreover, there were no correlations between the proportions of statements concerning acceleration and the total excessive acceleration score (i.e. area above 50% and below the time/throttle position curve: Spearman's $\rho = 0.12$, $p = 0.623$) or between acceleration codes and *total* acceleration (i.e. the total area under the time/throttle position curve: Spearman's $\rho = 0.095$, $p = 0.699$).

Each transcript was inspected for the raw total number of times each participant used the words 'accelerate', 'acceleration', or 'accelerating' in reference to their own behaviour (instances of avoiding the accelerator, or the use of 'accelerate' when referring to other drivers' behaviour, were not included). This revealed great variation between participants, with 7 of the 19 participants not referring to acceleration in this way at all, and 1 participant doing so 16 times (M = 2.47, SD = 4.01). There was no correlation whatsoever between excessive acceleration (in the vehicle data) and reference to 'acceleration' in the transcripts (Spearman's ρ = −0.077, p = 0.755). Looking at the total acceleration usage (i.e. the total area under the curve of throttle position by time) revealed a relationship in the *opposite* direction to that which might be expected, i.e. more references in the transcripts related to lower acceleration usage in the vehicle data; however, this trend was not significant (Spearman's ρ = −0.397, p = 0.093).

With regard to those participants that exhibited both higher coasting behaviours *and* lower excessive acceleration, it can be seen from Figure 4.5 that, if anything, the differences are even smaller than those based on other group separations (Figures 4.2 through 4.4), with no patterns emerging whatsoever. Although we have argued above that these five participants exhibited behaviours indicative of a fuel-efficient driving style, there appear to be no noticeable differences in their transcripts compared to the remainder of the participants.

In a final attempt to identify any possible differences in verbal reports between participants (based on the groups created by excessive acceleration and proportion coasting, and those participants that exhibited high coasting *and* low excessive acceleration compared to others), Mann–Whitney U tests were performed for every subcode (excluding the 'other' category), for all three grouping variables. This resulted in the performance of 111 separate tests; hence one would expect, by chance, around 5 of these tests to return a statistically significant result. In fact only two statistically significant results were observed; the low coasting group produced more statements under the code 'act to maintain speed' (code 1.3.3 in Table 1; M = 2.72, SD = 1.40) than did the high coasting group (M = 0.918, SD = 1.10; Mann–Whitney U = 13.0, p = 0.008), and the low excessive acceleration group produced fewer statements under 'anticipate behaviour of others' (code 5.3 in Table 1; M = 0.546, SD = 0.783) than did the high excessive acceleration group (M = 1.66, SD = 0.945; Mann–Whitney U = 12.0, p = 0.007). We therefore tentatively conclude that there are no differences between the verbal reports of those drivers who display more efficient driving behaviours and those who exhibit less efficient driving behaviours, at least with respect to harsh acceleration and the use of the vehicle's momentum.

4.5 DISCUSSION

As would be expected from any investigation of on-road behaviour, different drivers performed differently. Some drove faster than others, and some were more likely to drive more aggressively than others. Although effort was made to ensure similar traffic conditions across participants, in an on-road environment it is of course impossible to hold all extraneous variables constant; however, using measures of proportion (e.g. proportion of time coasting), rather than gross figures (e.g. total time spent coasting), should minimise the effect of varying traffic conditions. These measures (namely proportion of acceleration spent above 50% throttle depression, and proportion of travel time spent without depressing the throttle) did indeed reveal differences between participants; some participants used the momentum of the vehicle over throttle depression for forward travel more so than others (argued to be an efficient strategy; e.g. Staubach et al. 2014; Wu et al. 2011), and some participants had a greater tendency towards harsher acceleration than others (argued to be an inefficient strategy; e.g. Birrell et al. 2013). Whether these differences are reflected in underlying cognitive processes and structures is, however, a question that cannot be confidently answered here. No relationships whatsoever were found between the objective, quantitative measures of behaviour obtained from the instrumented vehicle, and the qualitative, subjective measures of cognitive processes obtained from the verbal reports, at least in terms of the group separations investigated here.

We, of course, do not conclude from this that there exists no relationship between cognition and action in driving, rather that the analysis of unguided verbal reports (i.e. in the sense argued for by Ericsson and Simon 1993) may be an inappropriate means for investigating such a relationship, at least when looking at groups whose differences may only be subtle, rather than clearly defined (e.g. novices versus experts (Lansdown 2002), or car drivers versus motorcyclists (Walker et al. 2011). The sample used in this chapter was not drawn from two separate populations; rather it was split into groups, in a *post hoc* fashion, based on objective vehicle data. Moreover, the group distinctions were not the same for the two variables used here; namely excessive acceleration and proportion coasting.

In the articles referenced in the introduction, group distinctions were clear (Lansdown 2002; Pampel et al. 2015; Walker et al. 2001a, 2011), either in terms of training and experience (novice/expert, driver/motorcyclist), or in equipment (high feedback/low feedback vehicles; though note that the two groups in Walker et al. (2001a) also differed significantly in age and experience). Moreover, research reported by Pampel et al. (2015) had participants drive in different ways for each trial, i.e. 'normally', 'safely', or 'efficiently'. They were, therefore, directed in their attention and behaviour (note also that Pampel et al. (2015) used a within-subjects design, therefore greatly reducing the compounding effect of individual differences). It may be that these a priori differences were necessary to show *post hoc* differences in verbal reports.

This could be linked to the fact that driving is, for most, a highly practised and therefore skilled activity, something that Hayes (1986) discusses in his review of Ericsson and Simon's thesis on verbal protocol analysis: 'Highly practiced processes may become so automated, however, that no intermediate products are available to STM [short-term memory]: The CP [central processor] may process the information without intermediate stores. This is especially obvious in the perceptual–motor area, but, significantly, even some verbal tasks can become automatic and thus not be fully available to STM. Further, only the products of cognitive processes are available to STM, not the processes per se' (Hayes 1986, 352).

This suggests that the processes themselves may not actually be available for verbalisations; rather, only the resulting *products* of cognition are available. Similar assertions were made by Pirolli and Recker (1994); they argued that performance driven by procedural knowledge [that which can be expected to dominate in highly practised tasks, as opposed to declarative knowledge, e.g. Anderson (1993)] is fast, automatic and unavailable for reflection by introspection. Results from our own previous work (McIlroy et al. 2012), in addition to earlier research from Isenberg (1986), supports this position. That automatic motor processes are not available to STM, hence are not available for verbalisation, is important for the purposes of understanding driving behaviours in an unguided study (i.e. without instruction or framing).

In studies where the participants understand, or guess the purpose of the research (or, indeed, are told), there is perhaps a subconscious bias towards framing their verbalisations in terms of their expectations of the study's desired outcomes. Nisbett and Wilson (1977) provide arguments pertinent to this point of view. They suggest that differences in verbal reports between, for example, experts and novices, may not be due to differences in underlying cognition per se, but that they arise from the knowledge the expert holds about what they *should* be focusing on or thinking about. It is that they remember the formal rules they were taught, and report on these as guiding behaviour, rather than reporting on the cognitive processes themselves. They offer the following point when talking about the weight an individual assigns to a particular piece of information in guiding behaviour: '... university admissions officials will be reasonably accurate about the weights they assign to various types of information in admissions folders, and auto mechanics will be reasonably accurate about the weights they assign to various factors in deciding whether a car has ignition or carburettor troubles. But such accuracy cannot be regarded as evidence of direct access to processes of evaluation. It is evidence for nothing more than the ability to describe the formal rules of evaluation' (Nisbett and Wilson 1977, 254).

In terms of driving, a similar argument applies; the processes themselves may be unavailable for introspection (due to them being automatic), hence any observed differences in verbal reports (e.g. between drivers and motorcyclists, or between people asked to drive 'normally' or 'efficiently') result from differences in that which they expect, or think *should* guide their behaviour, based on previous training or experimental instruction. This perspective is lent support from a comment made by Pampel et al.: 'the downside of the method [verbal protocol analysis] is the incompleteness of the verbalisations and in some places a mismatch between what the drivers said they intended to do and the behavioural data' (Pampel et al. 2015, 678).

It is quite conceivable that the differences in verbal reports shown by Pampel et al. (2015) were not only (or, perhaps, even) a result of differences in behaviour and cognition, but differences in what participants *think* efficient driving to be compared to 'normal' driving. That they observed no differences between normal and safe driving further bolsters this conclusion. Currently in the UK, when training people to drive significant attention is given to safety, with little focus on efficiency. Hence, 'normal' driving can be said to equate to 'safe' driving, as is it 'safe' driving that people have been trained to perform. Only when asked to perform 'efficient' driving do expectations or motivations change, thereby giving rise to differences in verbal reports.

In this chapter the participants were not asked to focus on any one aspect, they were not instructed to drive in a particular fashion, all had received similar training (i.e. licensed in Australia, with no advanced training), and all had a similar amount of experience on the roads. Furthermore, the eco-driving perspective adopted in this chapter was not mentioned. Without such differences, the only formal rules to which the participants had recourse (on which they could verbalise) were those taught in driver training; a training process shared by all.

At this point it is perhaps useful to draw on both Neisser's perceptual cycle model (PCM; Neisser 1976) and Rasmussen's skills, rules, and knowledge taxonomy (SRK; Rasmussen 1983), and to discuss the use of top-down and bottom-up processing, to further explain why it might be that Pampel et al.'s participants showed differences only when asked to drive economically, and why our participants showed no differences whatsoever.

The PCM is a cyclical information-processing model that suggests that environmental conditions trigger mental representations of the world (schemata), that these schemata guide our behaviour, and that in turn, our behaviour in the world (including perceptual exploration of the environment) modifies and updates our schemata (thus continuing the cycle). As described in Chapters 1 and 2 of this book, the SRK taxonomy distinguishes between three levels of cognitive control under which an individual interacts with the environment; skill-based behaviour (SBB) refers to automatic behaviour requiring little (if any) conscious monitoring; rule-based behaviour (RBB) encompasses behaviours driven by the associations made between familiar perceptual cues in the environment and stored rules for action; knowledge-based behaviour (KBB) requires effortful processing and analytical problem-solving based on symbolic reasoning and stored mental representations of the world (akin to schemata).

The distinction between top-down and bottom-up processing is one that was made explicitly by Neisser in the original PCM model (Neisser 1976), and is a distinction that can be inferred from the SRK taxonomy, inasmuch as KBB is characterised by top-down processing (using stored mental representations for complex reasoning), and SBB and RBB are characterised by bottom-up processing (more immediate reactions to information and stimuli in the environment). In this chapter's results a large proportion of the statements produced by the participants came under the following five subcodes; 'directional', 'general search', 'describe fixed environment', 'describe temporary environment', and 'describe other vehicles'. These are all related to gathering or describing information from or in the world; they are not reflective of schemata. Moreover, 56% of statements were categorised under the broader 'state' group. This is suggestive of a world-driven, bottom-up form of processing; this is, perhaps, unsurprising given the skilled, almost automatic nature of the driving task (as aforementioned). It is possible, therefore, that differences in cognitive processes were not revealed, as the mental representations in which we are interested are not the primary driving force of behaviour in this context, rather 'the world' is, hence the high proportion of statements relating to the external environment.

As Plant and Stanton (2015) discussed, in the PCM the use of general knowledge, and of the characteristics of the environment and the goals to be achieved, manifests itself as top-down processing and the use of schemata; or, in other words, behaviour at the knowledge-based level of cognitive control. Participants in Pampel et al.'s (2015) research showed differences under conditions of 'efficient' driving as this was something that could not be driven in a bottom-up way, at the skill level of cognitive control, hence may have been driven to a greater extent by schemata and, therefore knowledge-based reasoning. 'Safe' driving, on the other hand, can be considered to be equal to 'normal' driving (see above). They are driven at the skill-based level of control; hence the lack of differences between these two conditions.

Finally, it is important to point out that the preceding arguments by no means indicate that we consider the verbal protocol analysis method to be of no use in human factors and ergonomics research. Rather, in a context such as this (i.e. for behaviour at or approaching automaticity, without additional instruction or guidance) it may not be suitable for the identification of subtle differences in behaviour or cognition. Although Nisbett and Wilson's (1977) arguments (presented above) were given as part of a refutation of the 'verbal reports as data' approach as a whole (Ericsson and Simon 1980, 1993), we do not agree with their sentiments entirely. Verbal reports can and do provide useful information regarding group differences; whether the differences arise from underlying cognition or from the recall of that which has been previously learned may sometimes be less important than the fact that differences have been observed.

For example, we can still learn from the fact that motorcyclists and car drivers have incompatibilities when it comes to road situation appraisal, as evidenced by verbal report data (Walker et al. 2011), regardless of whether those verbal reports truly reflect underlying cognitive processes or not. It may be that cognitive processes are *not* available to the individual, and that they are reporting on rules learned in training; or it may be that previous training has changed cognition, and the drivers *are* reporting on cognitive processes. Whichever is the case, the differences are still of interest. The same applies to results from Pampel et al. (2015); whether cognitive processes change when asking people to drive efficiently, or whether they simply report that which they think they *should* (given the instruction), the differences in verbal reports still inform us of the knowledge people hold of eco-driving strategies, and of their mental models (Pampel et al. 2015).

4.6 CONCLUSIONS

This chapter has presented an analysis of the unguided verbal reports of drivers in an on-road setting, aiming to identify possible differences in underlying cognitive processes between those who drive more or less efficiently, as measured by the harshness of acceleration and the amount of time spent travelling without depressing the accelerator pedal. Although differences between groups of drivers have, in the past, been reflected in comments made while thinking aloud, this study has shown no such differences; no relationships could be found between quantitative vehicle data and qualitative verbal report data. This suggests that verbal protocol analysis may not be suited to the identification of subtle differences between drivers' cognitive strategies or processes, particularly when there is no clear separation between groups, or when there are no a priori instructions that specify the nature of the study or the style in which a participant is required to drive.

In terms of the main practical aim of this book as a whole, namely the support of eco-driving in the vehicle, it is difficult to see how the results presented above could go on to inform the design of an in-vehicle system that helps the driver to maximise efficiency and take full advantage of his or her vehicle's potential range. As with any research effort, the results cannot be known at the outset of the project; if this were the case, there would be no need to conduct the research. That being said, before undertaking the analysis presented in this chapter, differences between drivers were expected to come out in the verbal reports. It is, perhaps, possible that a different design may have given rise to more easily detectable differences between participants. For example, if the participants had been primed about the eco-driving focus given to the analysis of the data, more comments concerning eco-driving strategies may have been made. Such an approach would not, however, been in line with the verbal protocol analysis technique as argued for by its originators (Ericsson and Simon 1980, 1993). As such, the methodological discussions made above would not have been possible.

A negative result in any scientific domain is often a hard thing to sell. The aim of most scientific endeavours is to discover the existence of some phenomenon; justifying research that finds *no* evidence of something is more difficult, as there are likely to also be alternative explanations for the lack of a finding, rather than simply that the phenomenon in question does not exist. Although we agree with a number of academics, that is, that negative results provide a valuable contribution to the scientific literature (see, e.g. Matosin et al. 2014), the results from this chapter have not provided any clear information that might help to design an in-vehicle, eco-driving information device. The following chapter therefore takes an entirely different approach, returning to the ecological interface design methodology introduced in Chapters 1 and 2. The reader will see in the latter part of this book that the method itself has not been used in its entirety (hence the system described and tested in Chapters 7 to 9 cannot be said to be an 'ecological interface' per se); however, the SRK taxonomy, a fundamental component of the method, does represent the focus of the analysis effort (Chapter 6), and did provide the theoretical justifications for the design of the system (Chapters 7 and 8). The following chapter therefore provides a detailed discussion of the original method, its previous applications and of the importance of the SRK taxonomy. The discussion in the following chapter therefore presents the first step on the journey from theory, to analysis (Chapter 6), to design (Chapter 7) and testing (Chapters 8 and 9).

5 Two Decades of Ecological Interface Design, and the Importance of the SRK Taxonomy

5.1 INTRODUCTION

In the previous two chapters the practice of eco-driving was investigated through the use of two distinct methodologies; an online survey of 321 respondents, and the verbal protocol analysis (Ericsson and Simon 1980, 1993) of 19 participants in an on-road setting. The first of these chapters was a general investigation of eco-driving as a concept, and of the perceptions, attitudes and knowledge the general public have of it; it was not specifically intended to inform in-vehicle information design. Chapter 4, on the other hand, *was* motivated by such a goal. In particular, it was hoped that revealing the cognitive strategies of those that exhibit behaviours characteristic of efficient driving would help guide the design of a system that supports such behaviours in those that do *not* spontaneously exhibit them. This goal was not attained; this chapter therefore moves away from looking at the cognition and behaviour of individual drivers, and returns to the design framework introduced earlier in the book, that is, Ecological Interface Design (EID).

The method was introduced in Chapter 2 as a potential means for guiding the design of in-vehicle information systems, particularly with regard to fuel-efficient driving aids. The reader will later discover that the full EID method (as described by the method's creators) has not been used for the design of the in-vehicle system described and tested in Chapters 7 to 9. Attention has been paid to only one of its three core principles, a principle that arises from the underlying theory, that is, the skills, rules, and knowledge (SRK) taxonomy (as will be discussed); however, it is important to discuss the method as a whole. The initial intention of this research project was to use the method in its entirety; hence a thorough review of the past two decades of its applications was performed. The majority of the current chapter is devoted to reporting this review; however, as the research project progressed, and the theory explored to a deeper extent, the intention to use the full method was largely abandoned (as the reader will discover in Chapter 6). Nevertheless, the review presents a significant milestone in the journey of this book; without such a review, with particular focus on the SRK taxonomy of human behaviour, the remainder of this book would have taken a considerably different path.

5.2 ECOLOGICAL INTERFACE DESIGN

As aforementioned, ecological interface design (EID), a design framework based largely on the tenets of Gibsonian ecological psychology (Gibson 1979), was first fully described in the academic literature in the late 1980s (Rasmussen and Vicente 1989). Though it was initially developed for large-scale operations of industrial systems (with Rasmussen's background being in nuclear power research) it has, over the past 25 years, been used across many different domains including, but not limited to, aviation (e.g. Beevis et al. 1998), power plant refrigeration control (e.g. Lehane et al. 2000), military mission planning (e.g. Lintern et al. 2002), network management in information technology (e.g. Burns et al. 2003), petrochemical processing (e.g. Jamieson and Vicente 2001), private road vehicles (e.g. Young and Birrell 2012), intensive care units (e.g. Effken et al. 2008), and manufacturing control (e.g. Upton and Doherty 2008).

The method brings together two conceptual tools developed at the Electronics Department of the Risø National Laboratory in Denmark, namely the abstraction hierarchy (AH; Rasmussen 1985) and the SRK taxonomy (Rasmussen 1983). The AH is a tool used to model work domains at various (most commonly five) levels of abstraction, from describing the system's functional purpose (i.e. its reason for existence) at the highest level, to the physical objects that comprise the system at the lowest level. It presents a functionally organised hierarchy of information, where each node can be considered in terms of its reason for existence (i.e. why), and its realisation (i.e. how; Figure 5.1). As Vicente (1999) explained, the AH is used for work domain analysis, *not* task analysis; the analysis is independent of any particular workers, automations, events, tasks or interfaces. The SRK taxonomy describes three different levels of cognitive control with which actors interact with their environment (Rasmussen 1983). Skill-based behaviour (SBB) involves automatic, direct interaction with the environment; rule-based behaviours (RBB) involves associating familiar perceptual cues in the environment with stored rules for action and intent; knowledge-based behaviour (KBB) involves analytical problem-solving based on symbolic reasoning and stored mental models (Vicente 2002).

The EID framework as a whole is characterised by three general principles (described in detail in Vicente and Rasmussen 1992) that each correspond to a particular level of cognitive control, bringing together the SRK taxonomy and the AH conceptual tools. The general intention is that an interface developed according to EID will support each of the three levels of behaviour, with an interface adhering to the following:

- Skill-based behaviour – supports interaction via time-space signals; the operator should be able to act directly on the display, and the structure of the displayed information should be isomorphic to the part-whole structure of movements.

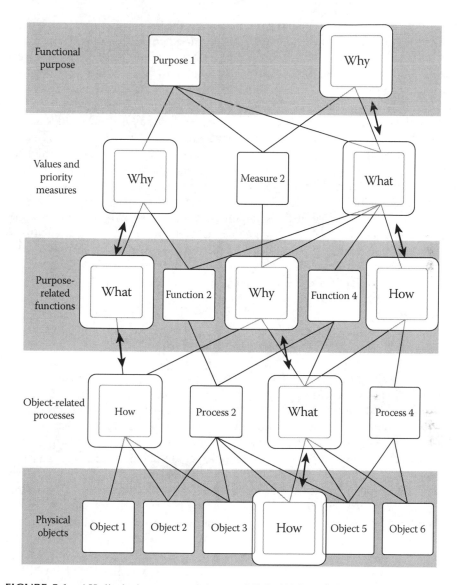

FIGURE 5.1 AH displaying means-ends causal links. (Adapted from McIlroy, R.C., and Stanton, N.A., *Theoretical Issues in Ergonomics Science*, *13*(4), 450–471, 2012.)

- Rule-based behaviour – provides a consistent one-to-one mapping between work domain constraints and the cues or signs provided by the interface.
- Knowledge-based behaviour – represents the work domain in the form of an AH to serve as an externalised mental model that will support knowledge-based reasoning.

Though support for all three levels of cognitive control is a necessary characteristic of EID, an equally important aim of the method is to provide an interface that supports, and indeed encourages reasoning at the lowest possible level of cognitive control, as it is at this level that behaviour is automatic (or at least approaching automatic), and it is most consistent, reliable and predictable (Rasmussen 1983). An interface should not force the operator to work at a higher level of cognitive control than the task demands. Another related goal of the methodology is to provide an interface that will support skill acquisition by encouraging the user to move through the levels of cognitive control, from KBB, through RBB, to SBB. This is achieved through the aggregation of individual actions (originally learned at the knowledge-based level and considered separately) and the higher-level cues, visual or otherwise, for those actions (characteristic of RBB), into more complex routines that are approaching automaticity (i.e. SBB) (Rasmussen and Vicente 1989).

Use of the term 'ecological' is important as it refers to the relationship the organism (or, in the case of human–machine systems, the worker) has with its environment, in terms of the ecology of the external world, the ecology of the system or interface, and the ecological validity of the cues in that interface [in the Brunswikian sense, i.e. the extent to which a cue in the interface maps on to, or correlates with, the distal variable in the environment to which it is related (Brunswik 1956)]. Gibson argued that a person can directly perceive the variables offered by their ecology (in this case the external environment) without the need for mediating information processing (Gibson 1979); EID aims to replicate this in an interface. By representing the work ecology in a faithful manner, providing ecologically valid cues and allowing for direct perception and manipulation of interface elements, it provides the user with a 'virtual ecology' (Rasmussen and Vicente 1989). If an interface is in keeping with the tenets of EID it will therefore reveal the goal-relevant constraints of the environment (i.e. the work ecology) in a manner that immediately reveals to the user the required goal-relevant actions and behaviours. In other words the interface provides a virtual ecology that maps the invariants of the work system such that the relevant affordances for action are revealed (Rasmussen et al. 1994). This removes the requirement on the user to create and maintain indirect mental representations of the system and the external reality, as the system will be represented in a way that requires minimal, if any, further processing to integrate information and infer the behaviour required by the system at any given point in time. Placing the need on the user to create indirect mental representations of a system should be avoided, as not only does this require more cognitive resources to construct, but the resulting models are also prone to inaccuracies and omissions (Gibson 1979; Revell and Stanton 2012).

5.3 COGNITIVE WORK ANALYSIS

Cognitive work analysis (CWA; e.g. Jenkins et al. 2009; Rasmussen et al. 1994; Vicente 1999) is an analysis technique that aims to describe the constraints acting on a work system at various levels of detail, from various perspectives, in order to show how a system *could* perform, rather than how it *should* or actually does perform (i.e. formative, rather than normative or descriptive). The CWA methodology is intimately linked with EID, being that they share two analytical tools (namely the AH and the SRK taxonomy), and a form of it was first described by Rasmussen in 1986 (note. though, that the specific term 'cognitive work analysis' was first used by Rasmussen, Pejtersen and Schmidt four years later in 1990).

The technique traditionally comprises five stages of analysis, each focusing on a different aspect of the system: work domain analysis, control task analysis, strategies analysis, social organisation and cooperation analysis, and worker competencies analysis. It is not necessarily required that all stages be performed; the technique may be considered more as a toolkit from which the human factors practitioner may use one or more parts, to suit her or his needs. The stages of the analysis chosen should reflect the constraints inherent in the system, as it is the constraints identified in analysis that guide the design of the system or interface. For example, whereas the social organisation and cooperation phase may not be of critical importance to an in-vehicle interface (where there will typically be only one primary user at a time), in an environment in which multiple actors are present, such as a hospital's operating theatre, the constraints arising from the different locations and information needs of the different actors will likely play an important role in the design of any information presentation system.

In terms of the original descriptions of EID (Rasmussen and Vicente 1989; Vicente and Rasmussen 1992), it is only the first and last stages of CWA, work domain analysis (WDA; in which the AH is developed) and worker competencies analysis (WCA; which uses the SRK taxonomy), which are discussed. As aforementioned, WDA describes the work domain in its entirety at various levels of abstraction, in order to display functional, means-ends relationships between system functions and components, which may be many-to-one or one-to-many as appropriate. The abstraction decomposition space can also be used in conjunction with the AH during this stage of the analysis (Vicente 1999). The abstraction decomposition space further specifies the WDA by decomposing the AH based on levels of resolution through the system; these part-whole decompositions often take the form of total system, subsystem, and components (see Vicente 1999).

The fifth stage of CWA, WCA, uses Rasmussen's (1983) SRK taxonomy to describe the level of cognitive control required by actors to fulfil different system functions. As described, when behaviour is skill-based (SBB), actions are automatic responses to environmental cues and events; little or no conscious effort is required. RBB relies on a set of rules and procedures held by the operator; control is characterised by these stored rules and procedures (as acquired through experience and formal or informal training). Here, specific goals need not be considered, rather behaviour is guided by if–then considerations, for example if stimulus x is recognised then response y is required. KBB requires slower, more effortful processing. This is used in instances where advanced reasoning is required; the user applies conscious attention and must carefully consider the functional principles that govern the system. This behaviour is most often seen in novel and unanticipated events and is more often exhibited by novice operators (Rasmussen 1983).

The other three phases of CWA were not part of the original description of the EID methodology (Rasmussen and Vicente 1989; Vicente and Rasmussen 1992);

however, they add significant value to a system analysis. The descriptions of the system provided by these phases, in terms of the control tasks, the strategies available to perform those tasks, and the social and organisational structure of the system (in terms of humans and/or technology), can contribute to interface design, hence a brief description of each will be offered here; for a more detailed discussion, the reader is referred to Vicente (1999) and Jenkins et al. (2009).

Control task analysis considers recurring activities in a system, focusing on what is to be achieved, independent of how the activity is to be conducted or by whom. It commonly represents system activity in terms of work functions and work situations; these situations may be spatial, temporal or a combination of both. It highlights situational constraints, describing when activities *can* be carried out, and when they are *likely* to be carried out. At least in more recent years, the contextual activity template (Naikar et al. 2006) has often used been in this phase of the analysis (Figure 5.2).

Also considered in this phase is an analysis of activity in decision-making terms; this is performed using Rasmussen's (1974) decision ladders. These diagrams capture the flow of information processing associated with the individual control tasks and are more common across the extant literature than the contextual activity template. Figure 5.3 displays an example of the decision ladder template; this will not be described in detail here, as a considerable amount of attention is paid to it in Chapter 6. It is sufficient to state here that the diagrams were initially intended to support design efforts, presenting graphically the cognitive processes performed by actors when undertaking a particular activity, in a particular context.

Strategies analysis addresses the constraints associated with the alternative ways with which each control task may be performed; it describes different methods of carrying out the same task. Some recent forms of strategies analysis have used representations such as Ahlstrom's flow diagrams (Ahlstrom 2005); these present

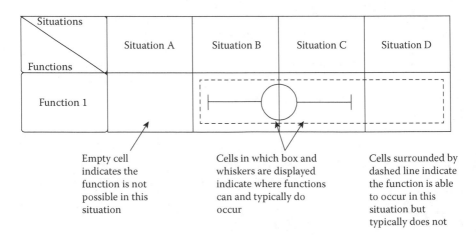

FIGURE 5.2 Explanatory figure of the Contextual Activity Template. (Adapted from Stanton, N.A., and McIlroy, R.C., *Theoretical Issues in Ergonomics Science*, *13*(2), 146–168, 2012.)

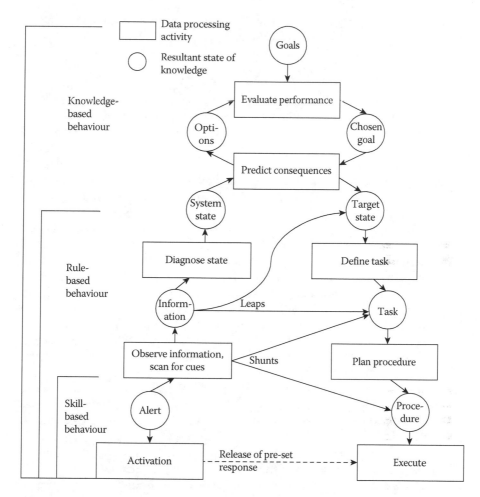

FIGURE 5.3 Decision ladder. (Adapted from Jenkins, D.P., et al., *Cognitive Work Analysis: Coping With Complexity*, Ashgate Publishing Limited, Farnham, England, 2009; Rasmussen, J., *The human data processor as a system component. Bits and pieces of a model. Riso-M-1722*, Roskilde, Denmark, 1974.)

a 'start state' and an 'end state' for a particular activity, with the two connected by a number of strategies, each describing different possible sequences of actions and operations.

Social organisation and cooperation analysis (SOCA) investigates the cooperation between actors in a system, addressing the constraints imposed by organisational structures and job roles and definitions. In this phase of CWA the contextual activity template can be used again; the representation developed in the control task analysis stage is coloured to indicate which actors (note that these can be human or technological) can perform the particular activity and in which situations. Decision ladders may also be used, as well as the flow diagrams used in strategies analysis;

indeed, the SOCA stage can be used to address any of the three prior analysis stages in terms of the actors involved in the system.

5.4   THE PAST 22 YEARS OF EID RESEARCH

Though early descriptions of EID only describe the use of the AH and the SRK taxonomy, it has been suggested that for the method to support integrated system design (rather than stand-alone interface design) it should expand its scope beyond that of work domain constraints and cognitive control levels (e.g. Vicente 2002). Rather than offering merely an analysis method, it has been suggested that CWA can be applied to work system design (e.g. Sanderson et al. 1999); hence some have argued for the use of all CWA phases when designing systems, with the resulting analysis going on to inform EID and overall system design, (e.g. Sanderson et al. 2000). Indeed, Rasmussen et al. (1994), in their seminal book *Cognitive Systems Engineering*, developed these concepts and further defined the contributions to system design of the various stages of CWA (though note that here the authors refer to a process that results in *ecological information systems* rather than *ecological interfaces* per se). Across the extant literature the usage of the different stages of CWA, at least in terms of informing EID, is far from consistent. Furthermore, the usage of the original form of the method in which only two sections are used, that is, the AH and the SRK taxonomy, also lacks consistency in the literature. A literature review of the past 22 years of journal articles and conference papers was therefore conducted; this timespan was chosen as it represents the EID research that has been conducted since the publication of Kim Vicente and Jens Rasmussen's *Ecological Interface Design: Theoretical Foundations* (Vicente and Rasmussen 1992), the first widely available journal article to describe and justify the method in detail.

Due to the high volume of research in CWA, particularly when considering any of the phases used alone or in combination, with or without explicit reference to CWA (e.g. see Read et al. 2012 for a review of the use of studies using the WDA phase for systems design), the current review was restricted to those papers explicitly referring to the EID method by name, (regardless of whether they mention CWA or not) and restricted to instances explicitly dealing with system design (i.e. some design products are presented), thus excluding those performing only an analysis of a system. It is important to clarify that this is not a review of studies using CWA, rather it is a review of EID literature.

Furthermore, this review represents an exploration of the literature detailing explicit, stand-alone applications of the method. Hence there are a number of texts exploring the theoretical foundations of the method that have not been included in this review. Two such examples stand out as providing excellent discussions of the method and its founding principles; Burns and Hajdukiewicz's book *Ecological Interface Design* (2004) and Bennett and Flach's (2011) more recent work *Display and Interface Design: Subtle Science, Exact Art*. Both texts provide thorough theoretical treatments of the method (particularly Bennett and Flach's), and both provide guidance for potential practitioners (particularly Burns and Hajdukiewicz's).

To identify appropriate articles, searches of relevant databases were undertaken. Databases searched included the publishers Taylor and Francis, Springerlink and Sage, the search engine Google Scholar, and Thomson Reuters' Web of Knowledge database. Only the term 'ecological interface design' was used (within quotation marks), as only papers explicitly referring to this method were of interest.

Following the search of databases, and the selection of articles that met the inclusion criteria (i.e. present design outputs rather than merely analysis), 75 entries were selected (Table 5.1). Note that the term 'entry' is used, rather than 'article' or 'report'; a number of the entries presented in the table are associated with more than one published article. This is due to the fact that in some instances a single interface design, and the analysis and design process therein, is described across multiple articles (e.g. from an ongoing research project, such as that of Amelink and colleagues; Amelink et al. 2003, 2005). This method of categorisation was chosen to avoid repetition, as this review intends to examine the different ways EID is used to develop separate interfaces, not the different descriptions of the same EID-guided interface and the preceding analyses. This method does, however, involve a degree of subjectivity. For example, there are instances where the same research group, working on the same overall system, have published work detailing the design of separate aspects of an interface for one system (e.g. in an airborne separation display; Van Dam et al. 2005; Ellerbroek et al. 2011, 2013). Although the analyses and subsequent interfaces presented in these three articles are for aspects of the same system, they have been judged as distinct enough, in terms of analysis and design, to merit their own entry into the table. Furthermore, two papers (Seppelt et al. 2005; Watson et al. 2000b) merited two entries. In Watson et al. (2000b) the authors deal with auditory displays designed according to EID guidelines. Here, two separate auditory interfaces are described, for entirely separate systems (one for anaesthesia monitoring and one for aircraft approach and landing). The report presented in Seppelt et al. (2005) is split into two sections; each section deals with the design of separate in-vehicle information systems.

It is recognised that this method of categorisation is not as objectively absolute as would be a method simply based on the numbers of published articles citing EID; however, as aforementioned the aim of this review is to deal with separate applications of the method rather than separate reports. This reveals more about how it is interpreted and used across different design applications, research institutes, and human factors and ergonomics practitioners, than would a review simply based on the number of papers citing EID (this would, e.g. incur a bias towards those researchers more disposed to extensively reporting their use of the method rather than reflecting the number of actual *applications* of the method to interface and system design). Moreover, as aforementioned this method of categorisation excludes important work on the theoretical development of the EID approach (e.g. Bennett and Flach 2011); however, this is primarily a review of EID applications, not an exploration of EID's theoretical underpinnings and advancements.

Before describing Table 5.1 it is necessary to note the differing use of terminology. Though all but three entries (Jungk et al. 1999, 2000 and Lindgren et al. 2009 cited as being based on the EID philosophy, each mentioning the SRK taxonomy, and Jamieson and Hilliard 2014, based on strategies analysis rather than WDA) at least mention the use of work domain analysis, the labelling of different analysis outputs (in terms of the representations of the work domain) differs across researchers. Nearly all specify the use of an AH, however, exactly what it is that is being referred to is not always the same.

Bisantz and Vicente (1994) describe the abstraction decomposition space (ADS) as an extension of the abstraction hierarchy (AH), with the part-whole decomposition (in the ADS) being conceptually orthogonal to the means-ends links described in the AH. The implication here is that the AH is developed first, followed by the separation of nodes across the part-whole dimension. This distinction is also implied in Vicente (1999). This being said, in Rasmussen's (1985) early description of the AH, the two concepts (functional decomposition and part-whole decomposition) are discussed with reference to only one diagram, and in Miller and Vicente (1998) it is implied that the two terms, 'abstraction hierarchy' and 'abstraction decomposition space', are interchangeable. To retain detail here, differentiation has been made between uses of the two decomposition dimensions. In the WDA column of Table 5.1, the presence of only 'AH' in a cell indicates that an article uses (or at least reports the use of) only the functional, means-ends abstraction dimension; the presence of the text 'AH/ADS' in a particular cell indicates where an article provides descriptions of both functional and part-whole decompositions. Only three other entries (across four articles; Cummings and Guerlain 2003; McEwen et al. 2012, 2014; Monta et al. 1999) do not specify the use of any particular representation for WDA, though all do cite the use of that analysis phase.

In summary, grey-shaded cells in the WDA column indicate that the analysis phase was mentioned by name; the addition of the text 'AH' represents use of an abstraction hierarchy using only functional abstraction, and the 'AH/ADS' text indicates use of both the functional and the part-whole dimensions. Grey shading in the control task analysis (ConTA) column indicates that control task analysis was referred to by name; the addition of the text 'DLs' indicates where decision ladder models were used for this phase. Grey shading in the strategies analysis (StrA) column indicates where strategies analysis was referred to by name; no formal outputs were provided for this stage by any entry, hence no text appears. In the SOCA column, grey shading represents where social organisation and cooperation analysis has been undertaken; the inclusion of the text 'DLs' in the entry for van Marwijk et al. (2011) indicates that these authors present decision ladders as the output of this phase. In the WCA column, grey shading without text indicates where the worker competencies analysis phase was referred to by name, but the SRK taxonomy was not cited (this is only the case for Effken, 2006). More commonly, the SRK taxonomy *is* cited and the WCA phase of CWA *is not* explicitly mentioned; this is denoted by the presence of the text 'SRK' without grey shading. Where both the text 'SRK' appears and the cell is shaded, both WCA and the SRK taxonomy are referred to by name. Finally, the CWA column is included to show where authors have also made reference to the CWA framework by name, as indicated by grey shading.

TABLE 5.1
Seventy-five EID Applications, Including Domain of Application, CWA Phases Discussed and Whether or not SRK Is Mentioned, Plus Notes

Authors and Year of Publication	Specific Domain	General Domain	WDA	ConTA	StrA	SOCA	WCA	Notes
Canfield and Petrucci (1993)	Computer-based patient records	Medicine	AH				SRK	Though a detailed analysis is not presented, it does describe the WDA and SRK principles that guide design.
Itoh et al. (1995); Naito et al. (1995)	Nuclear power plant control HMI	Power generation	AH/ADS				SRK	Naito et al. (1995) mention CWA, including control requirements, decision-making tasks, mental strategies and task allocation. However, analyses are not described or presented so one cannot assume that these authors carried them out (only that they have stated that this is what Rasmussen suggested).
Dinadis and Vicente (1995, 1996)	Feedwater subsystem	Power generation	AH/ADS				SRK	These papers present the first application of EID to a large-scale system, namely the interface for a complete feedwater subsystem of a nuclear power plant.
Vicente (1996)	DURESS and DURESS II	Thermal process control	AH/ADS				SRK	Provides an overview of work on DURESS and DURESS II up to this point. DURESS (DUal REServoir System Simulation); describes the first application of EID by the originators of the method (Vicente and Rasmussen 1990).
Beevis et al. (1998)	Aircraft engineering system	Aviation	AH/ADS				SRK	This research resulted in a prototype interface for the CC-130 Hercules aircraft engineering systems which was later evaluated in a focus group; it was concluded that EID was useful but needed to be supplemented by more specific design principles.
Sharp and Helmicki (1998)	Decision support in neonatal intensive care	Medicine	AH					These authors stated they used 'ecological interface design techniques' (350), namely the AH, citing the issue of clinical sensor technologies, defined a priori, as limiting the ability to measure variables defined in the AH.

(Continued)

TABLE 5.1 (Continued)
Seventy-five EID Applications, Including Domain of Application, CWA Phases Discussed and Whether or not SRK Is Mentioned, Plus Notes

Authors and Year of Publication	Specific Domain	General Domain	WDA	ConTA	StrA	SOCA	WCA	Notes
Chery et al. (1999)	Helicopter control display unit	Aviation	AH				SRK	The design and analysis process is not described in detail, with minimal elaboration on the utility of the SRK taxonomy.
Dinadis and Vicente (1999)	Aircraft systems display	Aviation	AH/ADS				SRK	This focuses on EID as more of a guiding philosophy than as a prescriptive design process.
Monta et al. (1999)	Supervisory control for water distribution	Resource distribution					SRK	Some graphical elements based on DURESS but this is vague. 'The implementation of an EID should be based on the results of careful and comprehensive WDA' (756) – but does not say how this went about it and does not present it.
Xu et al. (1999)	Hypertext (IT)	IT	AH/ADS					The interface developed is based solely on the AH, describing the AH 'as a semantic representation for an interface' that 'will support search and problem-solving activities' (207).
Jungk et al. (1999, 2000)	Haemodynamic monitoring	Medicine					SRK	Based on the *philosophy* of EID, rather than EID as a design procedure. Talks of different levels of abstraction to enhance KBB, though does not mention WDA specifically. Interfaces were partly based on Thull and Rau (1997), though this paper does not describe the process.
Burns (1999, 2000a, 2000b, 2000c)	Simulated coal power plant	Power generation	AH/ADS					(2000b): '10 views [in the interface] were designed to demonstrate the information in each cell of the AH' (114). (1999): interfaces are described and evaluated experimentally, but not actually shown. (2000c): provides more detail, but does not explicitly refer to EID, only 'ecological displays'.

(Continued)

TABLE 5.1 (Continued)

Seventy-five EID Applications, Including Domain of Application, CWA Phases Discussed and Whether or not SRK Is Mentioned, Plus Notes

Authors and Year of Publication	Specific Domain	General Domain	WDA	ConTA	StrA	SOCA	WCA	Notes
Lehane et al. (2000)	Refrigeration plant control	Thermal process control	AH				SRK	Cognitive task analysis is also reported; this analysed the schema used by the operator and thus 'determined the allocation of tasks between the stability program and the operator' (43).
Sawaragi et al. (2000)	Mobile robot teleoperation	Robotics	AH				SRK	Interface based on the concepts of EID (drawn from the SRK and the AH), but not based on any presented analyses.
Watson et al. (2000a) Watson et al. (2000b)	Anaesthesia	Medicine	AH					This was for an auditory display so no display actually presented, only described (being an auditory display it cannot be presented in a paper; this is the case for both papers).
Watson et al. (2000)	Approach and landing	Aviation	AH					Very basic design output in the form of a description of an auditory display (again, the output was not a visually presentable display). The phases are stated as being used, though only a basic AH is presented.
Ham and Yoon (2001a, 2001b)	Nuclear power	Power generation	AH				SRK	Ham and Yoon (2001b): HTA is used to analyse a typical task, though ConTA is not mentioned and it is not entirely clear how the HTA informed design. Ham and Yoon (2001a) provides another description of the same system, with a slightly different experimental evaluation focus.
Jamieson et al. (2001)	Acetylene hydrogenation reactor	Petrochemical	AH/ADS	DLs				HTA also used to identify requirements for known events. Jamieson and Ho (2001) describe the interface itself (not how it was developed). Jamieson (2002) provides an experimental evaluation of the interfaces.

(Continued)

TABLE 5.1 (Continued)

Seventy-five EID Applications, Including Domain of Application, CWA Phases Discussed and Whether or not SRK Is Mentioned, Plus Notes

Authors and Year of Publication	Specific Domain	General Domain	WDA	ConTA	StrA	SOCA	WCA	Notes
Jamieson and Vicente (2001)	Fluid catalytic cracking unit	Petrochemical	AH/ADS				SRK	This paper pays particular attention to the ability of an EID interface to support adaptive and continuous learning. The interface itself is not graphically presented, only described.
Lintern et al. (2002)	USAF mission planning	Military	AH/ADS	DLs			SRK	This short conference paper describes all CWA stages as feeding into design, though provides very little detail.
Reising and Sanderson (2002a)	Pasteurisation simulation	Pasteurisation	AH/ADS				SRK	The design is partly based on a WDA presented in a separate paper by the same authors (2002b).
Burns et al. (2003)	Oil extraction and upgrading facility	Petrochemical	AH/ADS					Also uses two additional information analysis techniques to improve the WDA/critical indicator analysis (what are the minimally necessary indications in the interface?) and contextual content analysis (adds richness to the data gathered so far). Only a perfunctory description of the analysis and design process is offered.
Burns et al. (2003)	Network management (IT)	IT	AH/ADS					In terms of actual design; 'For graphical visualizations of the variables [from the WDA], we adapted some previously tested and successful visual techniques from other domains' (376), i.e. EID did not support the whole process.
Cummings and Guerlain (2003)	Naval missile system	Military						All CWA phases mentioned, but no specific descriptions of how they proceed, or what representations are used, therefore cannot be assumed to have informed design.
Cummings et al. (2004)	Missile command and control	Military	AH					No SRK mention, though a very brief mention of KBB is provided; 'The means-end scheme represented through an AH provides a framework for highlighting areas which will require knowledge-based reasoning' (493).

(Continued)

TABLE 5.1 (Continued)
Seventy-five EID Applications, Including Domain of Application, CWA Phases Discussed and Whether or not SRK Is Mentioned, Plus Notes

Authors and Year of Publication	Specific Domain	General Domain	WDA	ConTA	StrA	SOCA	WCA	Notes
Dainoff et al. (2004)	Commercial investment software	IT	AH				SRK	Focuses more on CWA as informing design (rather than an EID focus), though states 'This display can be considered an example of an ecological interface since it supports: skill-based behavior by allowing direct manipulation of the interface, rule-based behavior by directly mapping the structure of the work domain, and knowledge-based behavior by representing the underlying constraints of the domain' (596–597).
Amelink et al. (2003, 2005)	Flight path display	Aviation	AH				SRK	SRK mentioned briefly in (2005) as a philosophy, not as a specific means for guiding actual design. See also (2003) for a conference paper detailing the same research.
Duez and Vicente (2005)	Network management (IT)	IT	AH				SRK	The AH is not presented, but the reader is referred to Duez (2003). See also Duez and Vicente (2003) for a conference paper detailing the same research.
Groskamp et al. (2005)	Anti-air warfare	Military	AH/ADS	DLs				Strategies are mentioned as routes on the ConTA decision ladder, though not in detail and not in relation to design.
Kruit et al. (2005)	Rally car drivers	Road transport	AH/ADS				SRK	This paper goes as far as to state that EID 'provides a prescriptive set of principles, regarding content, structure and form of an interface' (2). Note that we would disagree with the use of the word 'prescriptive' here.
Kwok and Burns (2005)	Mobile diabetes management display	Medicine	AH					Mentions skill-based reasoning but not the SRK taxonomy. The WDA is from 2003.
Memisevic et al. (2005)	Hydropower control	Power generation	AH/ADS				SRK	Brief mention of SRK, but no detail on how this informs design.

(Continued)

TABLE 5.1 (Continued)

Seventy-five EID Applications, Including Domain of Application, CWA Phases Discussed and Whether or not SRK Is Mentioned, Plus Notes

Authors and Year of Publication	Specific Domain	General Domain	WDA	ConTA	StrA	SOCA	WCA	Notes
Van Dam et al. (2005)	Airborne separation	Aviation	AH/ADS					See also Van Dam et al. (2004) – this is a less detailed report of the same research.
Borst et al. (2006)	Terrain warning system	Aviation	AH				SRK	Many of the constraints used in this study were identified in Amelink et al. (2005); this provides an analysis of the domain, but no design output.
Davies et al. (2006)	Sonar device for the visually impaired	Medicine	AH					A prototype auditory display is developed using EID and described (being auditory it cannot be visually presented). See also Davies et al. (2007) for prototype testing results.
Effken (2006)	ICU (medicine)	Medicine	AH (four levels)	DLs				Though 'skill level' is mentioned in terms of WCA, SRK is not cited. Builds on work from Effken et al. (2002); Effken, Loeb, Johnson, Johnson and Reyna (2002). See also Effken et al. (2008) for an experimental evaluation of interfaces.
Enomoto et al. (2006)	Cardiac nurse consultation	Medicine	AH/ADS					There is mention of rule- and knowledge-based behaviours, but no mention of SBB or the SRK more generally. The WDA is partly based on a strategies analysis (though not a traditional CWA-based StrA), which is presented in Burns et al. (2006).
Lau and Jamieson (2006)	Condenser subsystem for nuclear power	Power generation	AH/ADS				SRK	Offers a version of EID very close to the original descriptions; the AH provides 'a formative work analysis that should lead to a psychologically relevant and physically accurate representation of the work domain' (2) while the SRK is used to map 'the identified parameters, constraints, and invariants onto perceptual forms that capitalize on innate human capabilities' (3).

(Continued)

TABLE 5.1 (Continued)

Seventy-five EID Applications, Including Domain of Application, CWA Phases Discussed and Whether or not SRK Is Mentioned, Plus Notes

Authors and Year of Publication	Specific Domain	General Domain	WDA	ConTA	StrA	SOCA	WCA	Notes
Seppelt et al. (2005); Lee et al. (2006)	Lane change assist tool	Road transport	AH/ADS	DLs			SRK	Lee et al. (2006) presents only an empirical evaluation of the interface; preceding analyses and design processes, on which the interface was based, are presented in Seppelt et al. (2005).
Linegang et al. (2006)	Naval UAV control	Military	AH/ADS					Interface based solely on the AH; the AH is said to inform organisation of information for lower nodes based on their connections to higher nodes.
Pinder et al. (2006)	Aircraft thrust and brake indicator	Aviation	AH					Interface only based on AH, but no design process described. Only: 'The work domain analysis (WDA), shown in Figure 2, was used to develop alternative embodiments of the invention and particularly the prototype interface. The WDA captured the complex role of the thrust and braking indicator/advisor' (105).
Upton and Doherty (2006b)	High volume manufacturing	Manufacturing	AH/ADS			SOCA	SRK	A description of the social organisation of the system is offered (the authors have considered it) but no analysis technique/ representation is offered, neither is its influence on design.
Horiguchi et al. (2007)	Weighing machine	Manufacturing	AH					States that the AH 'can give designers a useful guide for organizing the display of all relevant information variables to be "externalized" in accordance with their means-end relationships' (885). Little more is given in terms of design procedure.
Jamieson et al. (2007)	Acetylene hydrogenation reactor	Petrochemical	AH/ADS	DLs			SRK	This also uses an HTA to bolster the information requirements garnered from the WDA.

(Continued)

TABLE 5.1 (Continued)
Seventy-five EID Applications, Including Domain of Application, CWA Phases Discussed and Whether or not SRK Is Mentioned, Plus Notes

Authors and Year of Publication	Specific Domain	General Domain	WDA	ConTA	StrA	SOCA	WCA	Notes
Seppelt et al. (2005) Seppelt and Lee (2007)	Adaptive cruise control display	Road transport	AH/ADS	DLs				Seppelt and Lee (2007) present only an empirical evaluation of the interface; preceding analyses and design processes, on which the interface was based, are presented in Seppelt et al. (2005).
Talcott et al. (2007); Bennett et al. (2008)	Army mission planning	Military	AH/ADS				SRK	An early version of the interface is presented in Talcott et al. (2007). The same interface, though a later version, is presented in Bennett et al. (2008). See also Hall et al. (2012); this also uses DLs in the analysis of the domain.
Upton and Doherty (2007)	Manufacturing process control	Manufacturing	AH					Task analysis methods not specifically described here. More details can be found in Upton and Doherty (2006a).
Van Dam et al. (2007, 2008a)	Airborne separation	Aviation	AH/ADS				SRK	In Van Dam et al. (2007) an extension of the interfaces presented in Van Dam et al. (2004, 2005) is provided. The AH/ADS is not mentioned or presented, but has been carried out in previous work. In Van Dam et al. (2008a) more detail is provided than in Van Dam et al. (2007). SRK is mentioned, but only after the design has been described – used to justify the design rather than to feed into it. See also Van Dam et al. (2008b) for an empirical evaluation of the interface.
Watson and Sanderson (2007)	Anaesthesia	Medicine	AH				SRK	This describes an auditory interface rather than presenting a visual one, providing a discussion of audition and the SRK taxonomy.

(Continued)

TABLE 5.1 (Continued)

Seventy-five EID Applications, Including Domain of Application, CWA Phases Discussed and Whether or not SRK Is Mentioned, Plus Notes

Authors and Year of Publication	Specific Domain	General Domain	WDA	ConTA	StrA	SOCA	WCA	Notes
Borst et al. (2008)	Aviation – terrain warning system	Aviation	AH				SRK	Largely based on the WDA from the Borst et al. (2006) work listed above; however, the interface design is different. The authors used DLs to model S-, R- and K-based behaviours for particular control tasks, though this is a post-design justification of the interface, not a method for informing design.
Hilliard and Jamieson (2008)	Solar vehicle	Road transport	AH					See also Hilliard and Jamieson (2007); here the display is described in more detail, though the analysis process is less detail.
Lau et al. (2008b)	Simulated BWR	Power generation	AH/ADS				SRK	See also Lau et al. (2008a) and Burns et al. (2008) for empirical evaluations of the developed interfaces.
Lee et al. (2008)	Private road vehicle	Road transport	AH				SRK	See also Nam and Myung (2007); this presents a more detailed description of the WDA, though does not present any design outputs.
Upton and Doherty (2008)	Process control health report	Manufacturing	AH/ADS				SRK	The AH/ADS (two of them) are from Upton and Doherty (2005). HTA is also used to inform design, and DLs for evaluation of designs. Like Upton and Doherty's previous work (2007) this uses the addition of task analysis to bolster information requirements.
Drivalou and Marmaras (2009); Drivalou (2005)	Electricity distribution	Resource distribution	AH				SRK	The interface is described in this paper; the analysis is from Drivalou (2005). See also Drivalou and Marmaras (2003, 2006) and Drivalou (2008) for descriptions of different aspects of the same research.

(Continued)

TABLE 5.1 (Continued)

Seventy-five EID Applications, Including Domain of Application, CWA Phases Discussed and Whether or not SRK Is Mentioned, Plus Notes

Authors and Year of Publication	Specific Domain	General Domain	WDA	ConTA	StrA	SOCA	WCA	Notes
Furukawa (2009)	Multiple robot supervision	Robotics	AH/ADS					See also Furukawa (2010); slightly less detailed description of the same research.
Lindgren et al. (2009)	Private road vehicle	Road transport					SRK	No description of design process, simply based on EID philosophy, mentioning SRK.
Morineau et al. (2009)	Tidal information display	Natural systems	AH/ADS				SRK	The authors acknowledge that 'our ED is not "a pure example of EID" (127), rather it is inspired by EID principles. Domain model not actually presented.
Gacias et al. (2010)	Vehicle routing	Road transport	AH					Knowledge-based reasoning is mentioned, but no SRK mention. StrA is cited, though unclear if it is undertaken. See also Gacias et al. (2009); here the problem is defined and a WDA is offered, though no resultant design is presented (this is only in Gacias et al. (2010)).
Horiguchi et al. (2010)	Robots in assembly operations	Manufacturing	AH					The authors also used HTA to specify target work domain; the HTA is not described as feeding into design.
Cleveland et al. (2011)	Air traffic control	Aviation	AH				SRK	States that an interface should encourage SBB and RBB while also supporting KBB; it does not, however, say how this is done. Minimal detail on the SRK framework is provided.
Ellerbroek et al. (2011)	Airborne separation	Aviation	AH				SRK	Related to Borst et al. (2008, 2006), Van Dam et al. (2007, 2008), but on a different aspect of flight path display, hence considered here as a new interface.
Jipp et al. (2011)	Airport management	Aviation	AH				SRK	AH is not actually presented, though each level is described in terms of the domain of interest.

(Continued)

TABLE 5.1 (Continued)

Seventy-five EID Applications, Including Domain of Application, CWA Phases Discussed and Whether or not SRK Is Mentioned, Plus Notes

Authors and Year of Publication	Specific Domain	General Domain	WDA	ConTA	StrA	SOCA	WCA	Notes
Mendoza et al. (2011)	Advanced driver assistance systems	Road transport	AH				SRK	Usability testing, heuristic evaluation and workshops are also used as part of a combination of EID and user-centred design.
van Marwijk et al. (2011)	Planning task in airborne separation	Aviation	AH/ADS	DLs		DL	SRK	Extension of Borst et al. (2008), Van Dam et al. (2007, 2008), but a different part of the interface is developed.
Kim et al. (2012)	Nuclear power	Power generation	AH				SRK	This paper only really used the AH, and only a three-level AH. There is no description of the usual functional means-ends links. SRK is mentioned as part of the EID concept, but no relation to this design or why it is important.
Lee (2012)	Nuclear power	Power generation	AH/ADS					EID is described as made up of the AH and the SRK taxonomy, though no detail on the design process is provided, and no description of what any part of the method aims to achieve. The authors also used cognitive task analysis (not ConTA specifically), though this is not explicitly linked to design.
McEwen et al. (2012, 2014)	Cardiac disease assessment	Medicine						Both papers describe a 'work analysis' that aims 'to discover the constraints of the work ecology, or in other words, to discover the deep structure of the problem' (McEwen et al. 2012, 2); however, no outputs are described or presented in either paper.

(Continued)

TABLE 5.1 (Continued)

Seventy-five EID Applications, Including Domain of Application, CWA Phases Discussed and Whether or not SRK Is Mentioned, Plus Notes

Authors and Year of Publication	Specific Domain	General Domain	WDA	ConTA	StrA	SOCA	WCA	Notes
Young and Birrell (2012)	Private road vehicle	Road transport	AH				SRK	Uses the AH presented in Birrell et al. (2008). The designs are based more on EID principles rather than using EID as a procedure for design – 'it was intended as a filtering stage between background analysis and more detailed interface evaluation, and as such makes no claims about the robustness of its scientific method', 235.
Ellerbroek et al. (2013)	Airborne separation	Aviation	AH					Closely related to Ellerbroek and colleagues' previous work (Ellerbroek et al. 2011). Based on the same analyses and for the same system; however, a different design concept is presented. Though SRK is not mentioned here, as in Ellerbroek et al. (2011) the study is said to employ a 'constraint-based approach, inspired by EID...'. See also (Ellerbroek, Brantegem, van Paassen, de Gelder and Mulder 2013) for an experimental evaluation of the interface.
Segall et al. (2013)	Decision support for anaesthesia	Medicine	AH					The authors used the WDA model presented in Hajdukiewicz et al. (2001).
van Paassen et al. (2013)	Air Traffic Management	Aviation	AH					Three partial AHs are presented, though no full AH. See also Klomp et al. (2012) for a conference paper detailing the same research. Furthermore, see Klomp et al. (2013) and Klomp et al. (2014) for experimental evaluations of, and further work on the resultant interface. SRK is not mentioned in any of the publications.

(Continued)

TABLE 5.1 (Continued)

Seventy-five EID Applications, Including Domain of Application, CWA Phases Discussed and Whether or not SRK Is Mentioned, Plus Notes

Authors and Year of Publication	Specific Domain	General Domain	WDA	ConTA	StrA	SOCA	WCA	Notes
Wright et al. (2013)	Information input (IT)	IT	AH				SRK	Recognises that the system state may be imperfect, arguing that the AH supports visibility of this overall state, thereby coping with imperfections.
Hilliard and Jamieson (2014)	Energy monitoring and targeting	Business processes					SRK	Explains EID as usually associated with WDA and SRK, though this takes a different approach, basing the design on StrA and the SRK taxonomy.
Li et al. (2014)	Physiotherapy assistant system	Medicine	AH/ADS				SRK	Both ADS and AH are presented, as is a justification of SRK usage. Though the three EID principles are not explicitly stated, it does state that SBB should be encouraged, and that all three levels should be supported.

ADS, abstraction decomposition space; AH, abstraction hierarchy; BWR, boiling water reactor; CAT, contextual activity template; ConTA, control task analysis; CWA, cognitive work analysis; DL, decision ladder; DURESS, dual reservoir system simulation; EID, ecological interface design; HTA, hierarchical task analysis; ICU, intensive care unit; IT, information technology; SOCA, social organisation and cooperation analysis; SRK, skills, rules and knowledge taxonomy; StrA, strategies analysis; UAV, unmanned air vehicle; USAF, United States Air Force; WCA, worker competencies analysis; WDA, work domain analysis.

5.5   EID APPLICATIONS

During early development of the EID method, applications were largely restricted
to process control (the first, and most extensively investigated interfaces of which
are the DURESS (DUal REservoir System Simulation) and DURESS II system
simulations, both of which are thermal-hydraulic process control microworlds;
Vicente 1996; Vicente and Rasmussen 1990). Its usage since these early studies has
spanned a number of distinctly different domains, the range of which is depicted
in Figure 5.4.

As can be seen from the graph, the aviation domain has received the most
attention in the EID literature, with 16 entries present in Table 5.1. Medicine is
second in terms of the number of entries presented in Table 5.1, with 12 applications falling under this heading. It is important to note here that these numbers
reflect distinct design applications, rather than simply citations of EID. Reports of
the experimental testing of EID interfaces are only included if they also present,

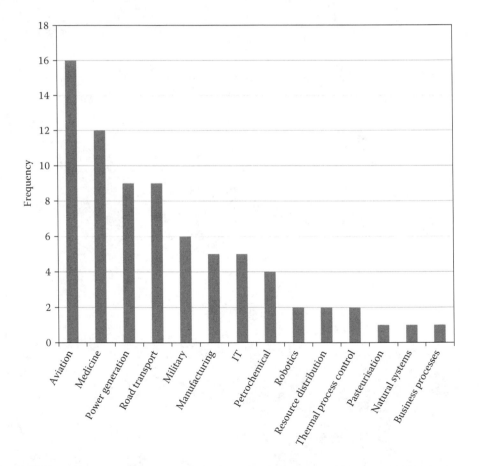

FIGURE 5.4   Domains of EID application, and their frequency, included in the current
review.

in the same article, a unique description of interface development, rather than using an interface already described in previous work. If such papers were to be included as separate entries, the thermal processes domain would feature more highly, as it is under this heading that studies related to DURESS and DURESS II fall. This simulation scenario was used as a test bed for the investigation of the EID concept, and there have been a considerable number of published studies assessing the effect of the DURESS interfaces on human behaviour (e.g. Carrasco et al. 2014; Christoffersen et al. 1994, 1998; Howie and Vicente 1998a, 1998b; Pawlak and Vicente 1996; St-Cyr et al. 2013; Torenvliet et al. 1998; Vicente 1997; Vicente et al. 1995, 1996).

Table 5.1 reveals the variation with which the different stages of EID and CWA have been employed, in various combinations, across the reviewed applications. This is graphically presented in Figure 5.5; the proportions of each of the combinations of the two original EID components, namely WDA and the SRK taxonomy, and the other CWA phases, namely ConTA, StrA, SOCA and WCA, are displayed. Note that for the purposes of clarity and detail, the SRK taxonomy is considered separately from CWA's WCA. Whereas WCA uses the SRK taxonomy, reference to SRK does not necessarily equate to a reference to WCA. In the early descriptions of EID (Rasmussen and Vicente 1989; Vicente and Rasmussen 1992) the SRK taxonomy is described as a stand-alone tool rather than part of an analysis of worker competencies.

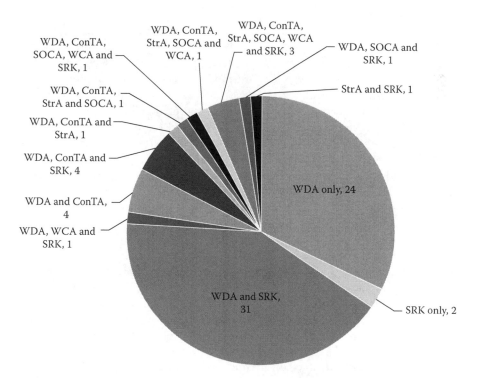

FIGURE 5.5 Proportions of the different combination of EID and CWA phases used across the reviewed literature.

Though it may be argued that an analysis of worker competencies implies usage of the SRK taxonomy (following Vicente's 1999 description of CWA), and that the use of the SRK taxonomy implies a consideration of the workers' competencies, usage and reporting styles in the reviewed literature do not necessarily make this link clear, suggesting that the relationship between SRK and WCA may not be consistently conceptualised as such across all researchers. Hence, in this review, an entry is only considered to have used WCA if the analysis step is explicitly referred to by name.

Though it is reasonable to assume there might be some differences in approach among various research institutes and individual researchers in the way they apply EID, and in the way it is reported (whether it is for a conference or journal paper, as an early prototype description or full-fledged interface design, across different domains and, indeed, simply based on a particular author's reporting style), the variation seen is indeed considerable. It is clear that the method has been interpreted and applied quite differently to the process described in early descriptions of EID. In the early work on EID by Vicente, Rasmussen and colleagues (Vicente 1996) descriptions of the design of the interface closely follow the original descriptions of EID, insofar as it is work domain analysis (WDA; either the AH or the further-defined ADS) and the SRK taxonomy that have guided design. The use of these two phases will therefore be explored in turn.

5.5.1 WORK DOMAIN ANALYSIS

As with many later papers, it is implied in Vicente (1996) that work domain analysis (WDA) has three primary functions in terms of design; (1) to offer an externalised model of the domain that serves to support knowledge-based reasoning (the KBB principle, see above); (2) to identify information requirements for an interface (i.e. the content of the interface); and (3) to inform the structure and organisation of the interface. Of the 75 entries presented in Table 5.1, 21 explicitly refer to the principle of supporting knowledge-based reasoning by providing an externalised model of the system (i.e. presenting the AH or ADS as part of the interface). Interestingly, two of these entries (Gacias et al. 2010; Xu et al. 1999) do not fully cite the SRK taxonomy; rather it is only the presentation of the AH as an externalised model to support users that is described.

In terms of points 2 and 3 described above, it is argued that a WDA (i.e. the AH or ADS) 'is used in EID to identify the information content and structure of the interface' (Vicente 1996, 252). As described earlier in this chapter, the WDA serves to describe the domain under analysis in a way that reveals functional links and constraints, and to provide a breakdown, at various levels of abstraction, of the individual functions and components of the system. It is the nodes presented in the hierarchy that inform *what* it is that needs to be displayed, and the means-ends links that provide information on how different aspects of the interface should link together. It informs the organisation of information displays such that the connections between separate physical components in the AH, and their links to higher, functional properties, are reflected in the organisation of the interface. Much of the focus of Burns and Hajdukiewicz's (2004) text on EID is on these two concepts; the provision of information requirements and the structure of the interface.

Vicente described these two benefits of WDA together; however, the two concepts are not always discussed together in the literature, with a number of researchers suggesting that WDA informs structure, with no reference to content, and a number of others stating content is informed, with no reference to structure. It is, however, difficult to clearly separate these instances, as often one is clearly stated and the other is implied, or both are implied in the same description of the method. Furthermore, as the WDA is considered such a fundamental part of EID (with all but two entries citing this stage; (Jungk et al. 1999, 2000; Mendoza et al. 2011) see above), many researchers do not go into considerable detail on the matter. Nevertheless, it is useful to highlight some evidence from across the literature explicitly referring to these concepts (Table 5.2).

TABLE 5.2
Evidence from the Literature Referring to the Utility of WDA in Informing Information Content and Interface Structure

Source	Evidence
Canfield and Petrucci (1993)	'the AH provides the foundation for interface design by specifying the information content and the structure of the interface' (3)
Sharp and Helmicki (1998)	'[WDA] captured the constraints that govern the behaviour of the system at various levels of abstraction and enable the identification of the variables important for system understanding' (3)
Burns (2000b)	'10 views [in the interface] were designed to demonstrate the information in each cell of the AH' (114)
Burns et al. (2003)	'from the model of how the system works, critical information for the displays can be identified' (370)
Cummings et al. (2004)	'the AH provided not only structure for domain information but also directly informed the design process' (493)
Memisevic et al. (2005)	Levels in the AH 'underscore the connection between functional properties and the physical components of the complex system' (3)
Seppelt et al. (2005)	'The AH identifies the information requirements for an interface through a systematic analysis of the drivers' environment' (10)
Horiguchi et al. (2007)	The AH 'can give designers a useful guide for organising the display of all relevant information variables' (885)
Bennett et al. (2008)	The AH and ADS 'provide a structured approach for a designer to determine the informational content that needs to be present in the display' (354)
Hilliard and Jamieson (2008)	The WDA reveals 'what needs to be presented' (7)
Lau et al. (2008b)	The AH/ADS reveals 'the constraints, invariants and parameters crucial to problem-solving that should be contained within an interface' (3582)
Lee et al. (2008)	The AH is about 'gathering these relationships in the form of information requirements' (388)

(Continued)

TABLE 5.2 (Continued)

Evidence from the Literature Referring to the Utility of WDA in Informing Information Content and Interface Structure

Source	Evidence
Horiguchi et al. (2010)	WDA reveal 'what kind of information should be specified for supporting quick and correct decisions' (2)
Cleveland et al. (2011)	The AH 'is the "map" that lays out the structure and constraints of the system' (209)
Young and Birrell (2012)	'The AH can be used to establish what kind of information is to be displayed, as well as where, when and how it should be presented, and finally how to integrate pieces of information which need to be associated' (227)
Segall et al. (2013)	'Means-ends relationships among variables [from the AH] are reflected in the grouping of data fields' (62)

5.5.2 SKILLS, RULES AND KNOWLEDGE

As described in early sections of this chapter, the skills, rules and knowledge (SRK) taxonomy represents the second of the two conceptual tools fundamental to EID. Where a model of the work domain is often said to provide information requirements (i.e. *what* needs to be presented; the *content*), it has been argued that the SRK taxonomy helps a designer with *how* that information is to be presented (i.e. the *form*). As Vicente put it, 'The SRK framework is used in EID to identify how information should be displayed in an interface. The idea is to take advantage of operators' powerful pattern recognition and psychomotor abilities' (Vicente 1996, 252). A central premise of EID is that performance in complex systems may be improved if an interface presents information such that it allows the operator to rely the skills they have developed through evolution (in terms of perceptual processing, pattern recognition and psychomotor abilities) when interacting with said interface (i.e. it encourages control at the skill- and rule-based levels). Table 5.3 presents some quotes from the literature referring to the utility of the SRK taxonomy in guiding the method of information presentation.

Despite the centrality of SRK to the EID method, its use across the reviewed literature is far from consistent. Of the 75 entries presented in Table 5.1, 45 make explicit reference to the SRK taxonomy, 30 do not. In other words, 40% of the entries citing EID as a guiding framework in their design methodology have not made any reference to the SRK taxonomy at all. It is worth noting that two thirds of the entries that do not cite SRK come from conference papers and other reports not published in peer-reviewed, academic journals (see Table 5.4). Of those appearing in peer-reviewed journals, 73.7% make reference to the SRK framework.

A two-tailed chi-square test of independence (with Yates's correction) was performed, revealing a significant relationship between presence or absence of SRK citation and the type of publication in which the article was found, χ^2 (1, $N = 75$) = 4.91, $p = 0.0267$. According to the odds ratio, articles that did not cite the SRK framework

TABLE 5.3

Evidence from the Literature Referring to the Guiding of Visual Form by the SRK Taxonomy

Source	Evidence
Itoh et al. (1995)	The SRK 'provides a basis for determining the form of information on the interface' (233)
Vicente (1996)	'The SRK framework is used in EID to identify how information should be displayed in an interface' (252)
Ham and Yoon (2001b)	The SRK 'provides a basis for determining the way to present the information' (104)
Drivalou (2005)	SRK-informed WCA 'provided insights for the visual form in which information should be displayed, in order to facilitate skill acquisition and support expert action' (265)
Seppelt et al. (2005)	'The SRK taxonomy of cognitive control provides guidance to the designer regarding how the format of information may influence how the driver is able to process information' (24)
Borst et al. (2006)	'the form that the interface will have is determined by the three levels of the skills, rules and knowledge taxonomy' (378)
Lau and Jamieson (2006)	'The domain characteristics captured in the WDA are mapped onto visual forms, as guided by the SRK taxonomy' (6)
Lau et al. (2008)	'The SRK taxonomy provides guidance on transforming information content and structure into perceptual forms by assisting designers in predicting the compatibility of representational forms with human information processing' (3585)
Li et al. (2014)	'The implementation of the skills, rules and knowledge (SRK) taxonomy, helped decide appropriate visualization forms for the functions extracted from the [WDA] models' (2)

TABLE 5.4

Contingency Table Displaying Presence or Absence of SRK Citation and the Type of Publication in Which the Article Appears

	Peer-reviewed Journal	Other Article Type	Total
Citing SRK	28	17	45
Not citing SRK	10	20	30
Total	38	37	75

were 3.29 times less likely to be found in peer-reviewed academic journals than those citing the framework.

5.5.3 REPORTED USE OF THE REMAINING CWA PHASES

Though only WDA and the SRK taxonomy are discussed in the early descriptions of EID, a number of other researchers have argued for the use of other CWA phases to enhance the EID process, both for interface design (e.g. van Marwijk et al. 2011; Watson et al. 2000a), and for system design more generally (e.g. Vicente 2002). The two methods are distinct; however, as aforementioned they are intimately related, and of the entries presented in Table 5.1, 21 explicitly make reference to CWA, either through referring to EID as a subset of CWA (e.g. Burns et al. 2003; Reising and Sanderson 2002a), by referring to EID as being based on CWA (e.g. Dainoff et al. 2004; Watson et al. 2000a), or simply by suggesting additional CWA phases may be used to bolster the analysis phase of EID regardless of actual application (Jamieson et al. 2007).

Figure 5.6 provides a graphical summary of the number of entries in which one or more of the remaining CWA phases, namely ConTA, StrA, SOCA and WCA, have been referred to as informing the design. Again, WCA has been separated from the SRK taxonomy. Figure 5.6 therefore only displays the number of entries explicitly citing 'worker competencies analysis'; entries citing the SRK taxonomy, but not referring to WCA by name, are not included. The number of entries citing all five CWA phases (including WDA) is also shown in Figure 5.6.

5.5.3.1 Control Task Analysis

As can be seen from Figure 5.6, control task analysis (ConTA) is the most commonly applied of the additional CWA phases, with 15 of the entries in Table 5.1 using this type of analysis. Of these 15 entries, 8 report the use of Rasmussen's decision ladders (Rasmussen 1974) and 7 do not report the use of any formal analysis output at all, hence none of the entries present the contextual activity template described by Naikar et al. (2006).

Incidentally, decision ladders draw on the SRK taxonomy to describe human control behaviours (a point discussed by Bennett and Flach 2011, and by Rasmussen 1974 in his original description of the models). Of the eight entries citing the use of decision ladders, only three also make any reference to the SRK taxonomy (Jamieson

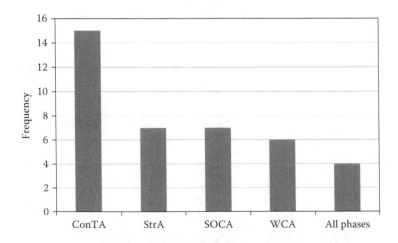

FIGURE 5.6 Frequencies of the usages of the different CWA phases.

et al. 2007; Lintern et al. 2002; van Marwijk et al. 2011), none of which do so in relation the decision ladders (i.e. the SRK is discussed separately).

In the reviewed literature, the direct link ConTA has to the design of the interface is often far from clear; however, a number of authors do provide some explanations. For example, Jamieson et al. (2001) argue that ConTA helps support management in predictable situations (where WDA is of more use in guiding performance in unanticipated, abnormal situations). They also argue that, in relation to task analysis, 'information requirements are deduced for the tasks and can serve as the basis for prioritizing, clustering, filtering, or sequencing information presentation elements in an interface design' (Jamieson et al. 2001, 10). The utility of ConTA in supporting the development of information requirements is again cited in later work by Jamieson and colleagues; 'work domain and task-based work analysis methods produce largely complementary IRs [information requirements]' (Jamieson et al. 2007, 904). A similar point is also made by Seppelt et al. (2005) in relation to the lane change advice system described in Lee et al. (2006); these authors argue that WDA and ConTA 'are both needed because the work domain and the control tasks both impose constraints on behaviour' (Seppelt et al. 2005, 19), suggesting that the outputs provide complementary information.

In Effken (2006) it is argued that the decision ladder outputs of ConTA describe the decision support at each step in the clinicians work that should be supported by the interface, as related to each level of the AH. Moreover it is suggested that ConTA can reveal which levels of abstraction are *not* used by clinicians, hence informs what does *not* need to be displayed (Effken 2006). Finally, both Upton and Doherty (2007) and Drivalou (2005) note the benefit of ConTA in helping to identify the different types of support required at different stages of task progression; ConTA informs 'the diverse types of operator support that is needed at different phases of operations' (Drivalou 2005, 265); task analysis 'showed that the different tasks require different perspectives on the display' (Upton and Doherty 2007, 177).

Finally, to return to the issue of content versus form, of all the entries detailing the link from ConTA to the design of the interface, only references to content are made. According to the reviewed literature, ConTA can be used to further inform *what* needs to be displayed in an interface; it does not guide *how* that information is to be displayed.

5.5.3.2 Strategies Analysis

From Figure 5.6 it can be seen that seven of the entries in Table 5.1 report the use of the strategies analysis (StrA) phase of CWA. All but one of these simply describes the process; six of the entries present no formal analysis outputs whatsoever. The exception is Seppelt et al. (2005) who use decision ladders for both ConTA and for StrA in their analysis of the Adaptive Cruise Control system (for the interface described in Seppelt and Lee 2007). Here it is argued that the StrA phase revealed transition points between tasks and the flow of information; these transition points provide context that dictates the required flow of information, further specifying the interface structure. Moreover, the authors state that the StrA identified gaps in system functioning, therefore identifying 'the ways in which the automation needed support through display design' (Seppelt et al. 2005, 55).

In Drivalou (2005) the description of StrA is slightly more detailed, with the author suggesting 'strategies analysis helped in identifying how the family of optional strategies that each control component affords for carrying out control tasks, influences the structure of operations. Different strategies have different sequence constraints; they have been used productively during the interface design, to define the different types of information representation needed to effectively support these strategies, as well as to easily navigate through the different displays and representations' (Drivalou 2005, 265).

Hilliard and Jamieson (2014) present a different approach to EID insofar as they use StrA, rather than a WDA, as the main input into design; hence in this paper the StrA process is described at length. The authors describe their information gathering exercises (literature reviews and a longitudinal study of domain practitioners), explain in detail the different strategies used in the domain, and describe how knowledge of these strategies was used to inform design. This paper presents an interesting approach to EID that as yet has received very little attention in the literature (though note Pejtersen's work on a library information retrieval system (Pejtersen 1989), work that is described in Rasmussen et al.'s seminal text *Cognitive Systems Engineering* (Rasmussen et al. 1994))

As with ConTA, where reference is made to the guiding of design, it is more often in relation to the information content of the interface rather than the form that information will (or should) take.

5.5.3.3   Social, Organisational and Cooperation Analysis

Of the entries presented in Table 5.1, seven report the use of social organisation and cooperation analysis (SOCA). In Upton and Doherty's work (2006b), though a description is offered of the social organisation of the workers, the article does not formally name SOCA. This would not be counted as using this analysis should the same criteria been applied as has been for WCA (i.e. it would have to have been explicitly referred to by name). It has, however, been included here as SOCA represents a completely distinct type of analysis (unlike WCA, in that it also uses the SRK taxonomy) and it is interesting to bring to attention where researchers have considered these organisational constraints when analysing a system.

In Watson et al. (2000), Watson and Sanderson (2007), and Effken (2006) the analysis step is reported as having been performed; however, no formal outputs and few details are provided. These simply state that the social organisation of the system was considered.

In Drivalou (2005), the link to design is made clearer. Here it is argued that 'Socio-organizational analysis helped in identifying how the affordances of available control artefacts structure the tasks at hand, and clarifying where task allocation is guided by pure organizational reasons and where is it guided by the capabilities of the artefacts' (Drivalou 2005, 265). It also states that 'making decisions about how work demands should be divided up had important implications for the identification and the definition of the relevant information content' (Drivalou 2005, 265).

Finally, van Marwijk et al.'s work (2011) is the only example from the entries presented in Table 5.1 of a paper to provide a formal analysis output from the SOCA phase. The authors offer two decision ladders that are said to represent 'the relationships between actions and subtasks of each actor in a system' (41). It is argued that this informs the allocation of function, particularly between pilot and automation (the domain of interest in their article). Furthermore, the authors argue that the interactions between the two decision ladders illustrate how the outcomes of one task (e.g. undertaken by the automation) can impact on the other (e.g. undertaken by the pilot), thus helping the designer to understand the effects of certain actions on the different system actors (van Marwijk et al. 2011).

5.5.3.4   Worker Competencies Analysis

As aforementioned, this phase of CWA uses the SRK taxonomy in its analysis of worker competencies; though many researchers cite the SRK taxonomy, explicit references to worker competencies analysis (WCA) are less common. From the entries presented in Table 5.1, six cite the use of WCA; none, however, provide formal procedural descriptions of the analysis stage, and none provide formal analysis outputs. In Reising and Sanderson (2002a) and Watson and Sanderson (2007) it is referred to in terms of skills-, rules- and knowledge-based behaviour representations. The descriptions of the analysis is similar in van Marwijk et al. (2011) in that it describes the purpose as being to 'identify the level of cognitive behaviour required to perform the tasks allocated to the human' (44), followed by a description of the three levels of cognitive control and how to support these behaviours. In Lintern et al. (2002) no information is given, other than that it should be carried out as part of the CWA process. Effken (2006) provides a little more information; here it is reported that the ICU is largely staffed by novices, and that this will affect the decision support required by the actors in the system (i.e. largely inexperienced actors). Of all the entries citing WCA, Drivalou (2005) provides the clearest link to design:

> Finally, we carried out competencies analysis. Through it we identified constraints related to the capabilities and limitations of the operators' cognition, and integrated them with requirements from previous phases. Competencies analysis also provided insights for the visual form in which information should be displayed, in order to facilitate skill acquisition and support expert action. (Drivalou 2005, 265)

Although there is no overt mention of the SRK taxonomy in Drivalou's (2005) description of the utility of WCA (though it is mentioned elsewhere in the article), it is interesting to see a reference to the guidance of visual form, a point noted by a number of other researchers in relation to said taxonomy (see Table 5.3). Furthermore, Drivalou (2005) refers to the notion of skill acquisition; this is another primary goal of EID (i.e. to foster learning) that is justified using the SRK taxonomy.

5.6   WHY THE SRK IS IMPORTANT

When following a particular human factors or ergonomics technique or method, and reporting its use in a published article, chapter, report or in a presentation or workshop, it is important to follow the method being cited (or at least explain and justify any deviations from that method). This may be an obvious point to make; however, if this is not done consistently, then the tools with which the human

factors and ergonomics community perform their work lose their scientific cred-
ibility; how can an author be said to be using a method if the process they report
differs considerably to, or omits sections of the method as originally described and
justified theoretically?

Differences in approach will of course be seen, and such differences are not
necessarily problematic. For example, it is not surprising that many of the research
applications reviewed here did not to include control task analysis, strategies analysis
or social organisation and cooperation analysis (or worker competencies analysis
specifically) when designing systems using EID; not only were these not part of
the original EID descriptions, these analyses can be time and resource intensive.
Depending on the goals and motivations guiding the analysis and design, these costs
may not be warranted. Furthermore, it can often be worthwhile to adapt a method
to suit the needs of the application; for example, in terms of EID and CWA, an
approach to analysis and design should be dependent on the constraints (to use a term
pertinent to CWA) that are important to a particular system or interface's design,
rather than some theoretical notions that are approaching a quarter of a century old.
It is important, however, to ensure that the manner in which a method is used, added
to or adapted is reported; this is at the fundament of academic research.

Whereas in some situations it may be useful or practicable to draw only on the
SRK taxonomy, or on work domain analysis alone to support interface design,
to report such applications as EID without explaining how it has been adapted,
or why this approach has been used rather than a form of EID more in line with
the original theoretical descriptions of the method, presents a worrying trend; the
literature reviewed here is rife with examples of this. In particular, it is in the variation with which the SRK taxonomy is cited in the EID literature that the issue lies,
as this is at the core of the original descriptions of EID; its three principles are
based on this very taxonomy.

In Rasmussen and Vicente (1989) the authors describe how it is the SRK taxonomy that provides EID with the theoretical grounding that separates it from previous design approaches, such as the direct manipulation interface approach (Hutchins
et al. 1986; Schneiderman 1983). The theory behind direct manipulation interfaces
(DMI) emphasises the need to represent objects of interest and to allow the users to
act directly on what they can see in the display (i.e. to combine action and control
surfaces). The DMI theory had significant influence on the EID framework; both
EID and DMI 'attempt to display the domain objects of interest and allow the operator to act directly on those objects' (Rasmussen and Vicente 1989, 527) and to allow
the operator 'to rely on the perceptual cues provided by the interface to control the
system' (525, ibid.). What separates the two approaches is the SRK taxonomy; where
DMI *describes* the direct manipulation principles, EID *explains* them in terms of
human processing capabilities (i.e. with the SRK taxonomy).

One explanation for the success of the DMI approach is that interfaces designed
using these principles foster feelings of direct engagement in the user; rather than
interacting with a mechanical or electronic intermediary, the user acts directly on
the concepts of the domain. Displaying to the user the relevant perceptual cues for
action is beneficial insofar as behaviour can be guided by a human's perceptual-
motor system (pertaining to skill- and rule-based reasoning), a system that is less

effortful and error-prone than the serial, analytical problem-solving based reasoning that characterises knowledge-based behaviour (Vicente 2002). The concept of designing an interface such that a user can act directly on its components leads on to the notion of combining action and observation surfaces; this is the coupling of the area on which a control action is performed (action surface) with the area from which system information is gathered (observation surface). Such an interface ensures 'that the time-space loop is maintained, thereby taking advantage of the efficiency of the human sensorimotor system' (Rasmussen and Vicente 1989, 528). This, again, is firmly rooted in the SRK philosophy; to encourage behaviour at lower levels of control will support more efficient, less effortful and less error-prone processing (though see Reason (1990) for a discussion on the different types of errors associated with differing levels of cognitive control).

Rasmussen and Vicente (1989; Vicente and Rasmussen 1992) describe the benefit of more rapid learning and skill development in an interface that combines both action and control surfaces. As has been described, this is another concept central to the EID philosophy; an interface should 'be designed in such a way that the aggregation of elementary movements into more complex routines corresponds with a concurrent integration (i.e. chunking) of visual features into higher level cues for these routines' (Rasmussen and Vicente 1989, 528). Such a description of the process of learning finds its basis in work from the field of psychology, with the taxonomy itself finding compatibility with earlier conceptualisations of human behaviour and learning, one of the earliest being that of Ryle's (1949) distinction of *knowing that* and *knowing how,* and the later, but closely related differentiation of declarative and procedural knowledge (Anderson 1976, 1983). Declarative knowledge (knowing that) refers to information in individual fragments that are stored separately, for example knowledge of facts, events and relationships, whereas procedural knowledge (knowing how) represents knowledge how to *do* things, for example complex motor skills, and cognitive skills and strategies. Where behaviour based on declarative knowledge requires effortful and time-consuming integration of knowledge fragments (Anderson 1993), with procedural knowledge the retrieval of information required to guide behaviour is said to be fast and automatic (Pirolli and Recker 1994). As Anderson (1993) explains, it is the conversion of declarative knowledge to procedural knowledge, through the amalgamation (or aggregation, in Rasmussen's words) of individual pieces of information into coherent concepts, or higher-level chunks that guide action, that characterises skill development, that is, learning.

These distinctions clearly resonate with the SRK philosophy; where knowledge-based behaviour requires the operator to perform complex reasoning, reflecting on and interpreting information displayed in the interface (using declarative knowledge), perceptual-motor reasoning (skill- and rule-based) needs only recognition of familiar aspects of the task or problem to guide behaviour (Glaser 1984). Such similarities between the SRK and earlier descriptions of human cognition are by no means accidental; Rasmussen et al. (1990) in an early description of CWA expressly state that the SRK taxonomy 'is compatible with the main-line [sic] of conceptualization within cognitive science and psychology (declarative vs. procedural knowledge ...)' (Rasmussen et al. 1990, 106).

Finally, there is a point to be made about the maturity of the method and the need to report SRK when using EID. Consider the following argument: human factors methods represent prescriptive techniques used to assist designers and engineers in their practices that are couched in descriptive accounts or theories of human behaviour and cognition. Once a method reaches maturity there need not be the requirement to explicitly cite the theoretical principles on which it was originally based. Though this argument may be valid in some instances (e.g. a detailed discussion of Ericsson and Simon's (1980, 1993) work on verbal protocol analysis may not be warranted every time a researcher applies the think-aloud procedure to elicit verbal reports from users or participants), we do not consider this to be the case for EID.

It is true that the SRK played an essential role in the development of the EID method, providing the fundamental descriptions of human control behaviour on which the method is based; however, we see it as more than simply a foundation, but as an integral part of the structure, both theoretical and practical, of EID. Where WDA provides a view on the system, the SRK provides a view on those who are to use that system.

Moreover, if a method is reaching maturity, one might expect it to be reported consistently in the literature; judging by the literature reviewed here this seems not to be the case. Additionally, there is no pattern of SRK citation frequency over time. One might expect there to be fewer references to the SRK framework as time progresses (i.e. as the method matures); however, this is also not the case. Up to the year 2000, 68% of entries in Table 5.1 referred to SRK; from 2001 to 2005, 53% did so; from 2006 to 2010, 55% did so; and since 2011, 67% of entries reported use of the SRK framework. Though it is possible that those researchers who fail to cite SRK still use it, and consider it as a founding principle of EID (but simply do not report doing so), the lack of a pattern over time suggests that its importance has not changed in the minds of EID practitioners.

This leads to the following question; *why* is the SRK is so often omitted? One explanation could simply be that WDA has been more often applied and reported in the literature than has the SRK taxonomy. This first step in CWA has received more attention than any of the other stages and the process for performing it is better defined, with more concrete analysis outputs (i.e. the AH and ADS) than SRK. Indeed, the SRK taxonomy is more of a description of human control behaviour than a prescriptive analysis technique.

This characteristic of the SRK taxonomy may be another contributing factor to its relative lack of citations. For example, take Bennett and Flach's (2011) book on EID and its theoretical underpinnings. In addition to a thorough and detailed discussion on the foundations of EID, a number of design tutorials are provided. In these examples the SRK concept is woven into the description of the process, rather than explained as a stand-alone tool (as is more the case for WDA). That its contribution to design is less easily described may contribute to the fact that it is sometimes omitted. Where WDA is a set procedure with a specific product (the AH or ADS), the SRK is a more general philosophy that affects the whole process, albeit in a less easily defined way.

In their design tutorials Bennett and Flach (2011) describe how the presented interface supports each of the levels of behavioural control; however, they do not explicitly say how the SRK taxonomy guided design a priori. This could suggest that the SRK taxonomy is not only a means for guiding design, but also a set of criteria against which a design must be judged. Note, however, that many of the concepts that are used to guide the form of the interface (e.g. the visual thesaurus, direct manipulation principles, configural graphics etc. (Bennett and Flach 2011)) themselves embed the SRK philosophy.

This concept also highlights a potential ambiguity regarding the direction of *how* interface elements are presented. In one sense, the SRK guides how information is to be presented in terms of the visual form the interface will take. In another, it describes how information should be displayed in terms of the level of cognitive control that will result from a given design. In Table 5.3 (above) the majority of researchers appear to be alluding to the first point, that is, the a priori guidance of visual form. Seppelt et al. (2005), on the other hand, appear to be referring more to the second point, that is, how the resulting interface will support skill-, rule- and knowledge-based behaviours. It is this conceptual approach that Bennett and Flach (2011) seem to favour in their design tutorials; however, note that a considerable portion of the work that precedes the design tutorials chapters deals with the former conceptualisation of SRK and its role in EID.

In Burns and Hajdukiewicz's seminal work *Ecological Interface Design* (2004) the SRK taxonomy is dealt with in an even less detailed manner. The term receives only three mentions, and all appear in the introduction (on pages 7, 9 and 10). Though it is stated that, according to EID, an interface or system should support all three levels of cognitive control, the three founding principles of EID are not provided, and although a small section is offered discussing the process of learning, it is not couched in terms of encouraging behaviour at the lowest level of cognitive control (a central tenet of EID according to its originators). The book does, however, provide a chapter on the creation of visual form, including a thorough discussion of a visual thesaurus (which could also be argued to embody the principles of EID, though this is not discussed in the book). Finally, in one of the design examples a point made about encouraging perceptually driven behaviour; 'forms were selected that would allow the practitioners to use their perceptual abilities as much as possible' (Burns and Hajdukiewicz 2004, 207). This is entirely in line with the SRK-driven EID theory (i.e. that SBB takes advantage of the human's perceptual processing abilities and hence should be encouraged); it is not, however, discussed in relation to, or justified by the taxonomy. This is by no means a criticism of Burns and Hajdukiewicz's work; they state quite clearly that the book is not intended as a theoretical treatment of the method, rather it is a pedagogical text aimed at students and practitioners (for theoretical discussions they direct the reader to Rasmussen and Vicente's original papers on EID).

Despite the potential difficulty with which an absolute, direct link between the SRK taxonomy and design can be drawn, it seems quite the omission to talk of EID in an academic journal article without mentioning the SRK taxonomy; if the process does not pay attention to the perceptual and cognitive capabilities of the human actor, it seems misplaced to use the term *ecological* interface design. It is, therefore argued here that a discussion of EID that does not recognise the influence of SRK on the methodology is a discussion that is incomplete.

5.7   CAN EID ALONE RESULT IN DESIGN?

Even with the inclusion of the SRK taxonomy (and the associated guidance on supporting human behaviour), an application of EID does not guarantee the creation of a successful design. Indeed, it could be argued that there is little way of confirming that the inclusion of any of the additional CWA stages necessarily results in a successful system design. Although many of the studies included in this review, and some that are not (i.e. separate articles presenting investigations of designs included in the current review), provide comparisons of an EID-based design with conventional, existing or simply non-EID designs, there exist no studies (to our knowledge) that compare interfaces developed with the different stages of CWA. Whether or not adding a stage of CWA to the pre-design analysis phase does indeed benefit the ultimate design solution is a question that cannot, as far as we are aware, be answered with data present in the extant literature.

The jump from analysis to design is by no means trivial, and a design method such as EID (or any other design methodology, for that matter) cannot definitively bridge this gap. Although the method does provide support to the design effort, a number of researchers have argued that it must be supplemented, and that when used in isolation, it may not even result in design at all (Table 5.5).

Reising and Sanderson (2002a) also make this point, suggesting that knowledge of *how* to display items in an interface largely comes from advice from other areas of ergonomics and psychology. Though the SRK taxonomy itself offers guidance in

TABLE 5.5
Quotes Alluding to the Inadequacy of EID When Used in Isolation

Source	Evidence
Dinadis and Vicente (1996)	'EID simply states that the constraints of the work domain should be mapped onto perceptual signs in the interface; it does not state how this is to be accomplished. Therefore, EID must be supplemented by more detailed design principles' (276)
Beevis et al. (1998)	'EID principles, in themselves, do not provide much detailed guidance on the implementation of the visual form of the necessary information' (2–5)
Seppelt et al. (2005)	The 'steps of the CWA process do not specify a particular interface. Instead, they indicate general constraints that a useful interface should consider' (25)
Borst et al. (2006)	'The framework incorporates guidelines to analyse the cognitive work domain, but does not include a specific recipe to determine what the interface should actually look like' (378)
Ellerbroek et al. (2011)	'It is clear that the step from WDA to display concept is far from a trivial one' (873)
Mendoza et al. (2011)	EID 'is more of a philosophical tool for the designer than a fully fledged method that can be applied without much effort. It is difficult to assess how the AH and SRK taxonomy is reflected in the interface design and how to be able to evaluate it' (58)

terms of leading a designer to think about the level of cognitive control of the potential actor, and the benefit of taking advantage of the human sensorimotor system (i.e. to support SBB), it says little of the fundamental human perceptual system itself. Dinadis and Vicente (1996) frame it thus:

> EID was only intended to address some basic issues in interface design, and so there are several important design problems for which it does not provide guidance. For example, the principle of visual momentum and perceptual organization principles had to be used to supplement the EID framework ... following the principles of EID alone does not allow one to design an effective interface for large scale systems. (Dinadis and Vicente 1996, 277)

Such considerations are not uncommon among those applying the EID method, be they implied or explicitly stated. A number of the articles reviewed above cite the use of additional tools external to those described as part of EID (or of CWA), ranging from knowledge of basic psychological principles of perception and cognition, to human factors and ergonomics analysis techniques and design heuristics. A number of examples of these additions are presented in Table 5.6. Describing

TABLE 5.6
Additions to the EID Process

Source	Additional Methods Used
Chery et al. (1999)	US Department of Defense design criteria
Lehane et al. (2000)	Cognitive task analysis
Watson et al. (2000a); Watson et al. (2000b)	Attentional mapping (for multimodal displays)
Ham and Yoon (2001b)	Hierarchical task analysis
Burns et al. (2003)	Critical indicator analysis; contextual content analysis
Cummings and Guerlain (2003)	Analysis of global organisation factors (similar to, but distinct from SOCA); creation of pilot domain (for prototyping)
Kwok and Burns (2005)	Hierarchical task analysis
Seppelt et al. (2005)	Salience effort expectancy value (SEEV) model (from Wickens et al. 2003); the semiology of graphics variables for graphic composition (from Bertin 1983)
Jamieson et al. (2007)	Hierarchical task analysis
Lau et al. (2008)	AH supplemented with causal links
Upton and Doherty (2008)	Hierarchical task analysis; the semiology of graphics variables for graphic composition (from Bertin 1983)
Drivalou and Marmaras (2009)	Spatial arrangement principles (from Wickens and Carswell 1995); configural display guidelines (from Hansen 1995)
Jipp et al. (2011)	Usability testing; heuristic evaluation; workshops
Cleveland et al. (2011)	Informal interviews
Kim et al. (2012)	Heuristic techniques (from Hansen 1995)
Lee (2012)	Cognitive task analysis
Young and Birrell (2012)	User Requirements Questionnaire
Segall et al. (2013)	Hierarchical task analysis; cognitive task analysis

all these supplements to EID and their benefits is outside the scope of this review, though see Burns and Hajdukiwicz for a full chapter on using EID with other methods (Burns and Hajdukiewicz 2004). The main point to make here is that though EID may not give a designer everything they need when designing an interface or system, the method is easily and, according to Dinadis and Vicente (1996), necessarily supplemented.

5.8   GENERAL DISCUSSION

The review presented in this chapter has revealed a significant amount of variation in the literature reporting EID as a guiding methodology, in terms of the usage of different analysis phases, its use across various domains, and even in the differential use of additional analysis techniques and design heuristics. It is therefore difficult to see how it would be possible to describe a strict procedure that guides an individual in their progress from concept formation, through analysis, to design. This is, however, by no means a criticism of EID as an approach to system design. It is merely an acceptance of the idea that simplifying the creativity and insight involved in interface and system design down to a formal, step-by-step procedure that will always result in a 'perfect' or 'ideal' interface is a goal that is (in our judgement) neither possible to attain nor useful to pursue.

There will always be a step from analysis to design; to think otherwise would be to wrongfully disregard the importance of creative thinking in design. The benefits of EID lie not in removing this step, but in supporting a detailed understanding of the system under development, and an appreciation for the cognitive capabilities and characteristics of those who are to use the system. Both of these are necessary if an interface is to contain the relevant information (presented at the relevant time) and display it in an appropriate manner.

Nevertheless, EID can also assist in the design of an interface's elements. Principles related to our knowledge of human behaviour (e.g. SRK, direct manipulation) and of human perception (e.g. visual thesauruses, configural graphics) can be applied to the design of perceptual form. Despite the nebulous nature of this design process (i.e. it cannot be easily defined as a prescriptive procedure), a discussion of the design activities is warranted when composing articles for publication in scientific journals. The issue is to be consistent and explicit in the reporting of the way the method has been used, and why it has been used in such a manner; this is at the very fundament of the scientific discipline.

Moreover, it may be that some minimum reporting requirements are warranted in EID research. WDA alone does not constitute an application of EID. It represents one aspect of EID (or of CWA), and this form of analysis has itself (in conjunction with other CWA phases) has been used to guide design without also referring to EID (see Read et al. (2012) for a review of CWA design applications). It is the inclusion of both WDA and the SRK taxonomy that signifies EID as a unique approach to design. Furthermore, it is possible that an interface can be described as being 'ecological' (i.e. typified by certain features such as configural graphics, or has 'ecological validity' inasmuch as cues in the interface are highly correlated with their referents in the external environment) without necessarily having been designed using EID. While it is not in the scope of this chapter to discuss the suitability of the term 'ecological' in relation to EID (the interested reader is referred to Bennett and Flach (2011)), the term 'ecological interface design' is the name of the method. Should an article cite use of the method, it would be appropriate to at least cite its core principles (i.e. WDA and SRK), even if these do not necessarily represent the main thrust of the research.

A further benefit to EID lies in its provision of a documented record of the analysis and design process. Following the EID (and, where appropriate, the additional CWA phases) approach to system analysis is not in merely about producing analysis outputs to hand over to a designer, which will then be transformed into an interface, rather it is in the process of analysis itself, and the structure that EID gives this process: 'much of the benefit to work analysis lies in the structured discovery process that it fosters' (Jamieson et al. 2007, 897).

Furthermore, the direction of focus that EID fosters also benefits the process insofar as it redirects the focus away from a physical component-based description of a system to an analysis of higher levels of system functions (Beevis et al. 1998). It supports a 'deep knowledge' of the set of constraints that characterise a system and its behaviour (Jamieson et al. 2001) and, through the SRK taxonomy, it leads the designer to think about the fundamental manner in which humans perceive and interact with their environment. These benefits, however, can only be properly taken advantage of if those who use the method are explicit in the way they use EID, either in its original form, or in an adapted or modified version.

The link between the various CWA phases and EID tools and the final design output could be made clearer; if an analysis step has been performed, then it should be reported, along with the outputs of this phase. Though it may not be practical in all cases to present the full analysis output (CWA often produces outputs too voluminous for inclusion in published articles), examples at least should be provided. Furthermore, if only a part of the method has been used, it may be useful to give a brief explanation of why this approach has been chosen over usage of the method in full. This will provide transparency in the literature, ensuring that other academics, researchers and designers can understand how that method was applied, why it was applied in that way, and how such an application can be repeated, adapted or built upon in subsequent research.

In addition, more detailed reporting will help clarify the contributions to design made by each specific CWA phase. The different phases each identify different types of system constraints and therefore (implicitly) a design opportunity. These constraints relate to the functions a system can perform with the current or potential configuration of physical objects (i.e. WDA), the situations that the system faces (i.e. ConTA), the decisions and strategies used (i.e. StrA), the allocation of functions to, and organisation of people and technology (i.e. SOCA) and the competencies of workers and technologies (i.e. WCA). Typically, when performing EID most only talk of WDA; little mention is made of the trade-off between the different sorts of constants acting on the design of interfaces. To deal with such trade-offs is no trivial problem, and indeed these other phases are not core components of EID as originally described. Note though that Vicente (2002), in his early review of EID applications, describes the benefits of expanding EID's scope beyond work domain and cognitive constraints to those covered by other CWA phases. Such a practice has potential to add great benefit to integrated system design; however, the methods and outputs of such a process must be reported if other researchers are to repeat and expand on the approach.

Differences in how EID supports the identification of required *content* and how it guides the *form* of the interface are also of interest. The general consensus in the literature reviewed here is that WDA guides the development of information requirements (i.e. content) and interface structure. This is also true for the StrA, ConTA and SOCA phases of CWA where applied; the literature suggests these inform interface content more so than form.

In terms of form, although both Dinadis and Vicente (1996) and Reising and Sanderson (2002a) suggest that to specify an interface's form requires further input from methods and principles external to EID, particularly in relation to the fundamental properties of the human perceptual system, a number of researchers do make reference to the SRK taxonomy informing how information should be presented (see Table 5.3). Where system analysis provides information requirements, the SRK, through providing a description of a human's processing tendencies and capabilities, helps a designer convert these requirements into forms to be presented to the end user. Though these forms are often visual, they could also be auditory, haptic or even olfactory, a point well illustrated in Watson and Sanderson's work (2007); here the SRK description of human cognition plays a crucial role in the design of, and indeed justification for auditory displays. Furthermore, though not an example of a design application (hence not included in Table 5.1), Lee et al. (2004) provide a theoretical discussion on EID and haptic interfaces; the implications for design are firmly rooted in the three SRK-based EID principles.

Finally, it is interesting to note that the majority of the entries in Table 5.1 that represent academic journal articles (i.e. those that have been through a rigorous peer-review process) do in fact cite the SRK taxonomy (73.7% compared to the 26.3% that omit any reference to SRK). The split is less one sided, and in the other direction, among the conference papers and other reports, where the peer-review process may be said to be less stringent; here, 45.9% do cite SRK, 54.1% do not. This may suggest that there is still general agreement among the ecological interface design community (the likely reviewers of EID related journal submissions) as to the importance of the SRK in EID research and application. However, that over a quarter of EID journal articles still do not make reference to SRK is still a statistic worthy of attention.

5.9   CONCLUSIONS

If EID is taken to include all phases of CWA, then it becomes a method that identifies successive constraints on the design of an interface or system, in terms of the work domain, the control tasks, the available strategies, the socio-technical organisation of the system, and the constraints relating to human's inherent cognitive processing tendencies. The constraints that are important to design depend on the type of system under development (as aforementioned, although it is still part of a larger traffic system, the user interface for a single road vehicle will likely be less impacted by social-organisational constraints than will, e.g. a hospital's operating theatre);

the choice of the analysis phases conducted should reflect this. The flexibility and adaptability of the design framework are important advantages, and its continued use is a testament to the utility of the method. Variation in approach over time, across researchers and between domains is natural and is not problematic; it is in the consistency and detail with which these applications are reported, and when and how the chosen approach differs from theoretical descriptions of the method, that progress in the field could be made. Across all scientific disciplines, documentation is critical; this is no different for the ergonomics and human factors domain.

We would like to emphasise once again the importance of the SRK taxonomy to the EID framework as a whole; the defining principles of EID are based on this description of human cognition. Constraints relating to these basic cognitive characteristics therefore necessarily affect the way in which humans use a given system. Citing EID without at the very least referring to the SRK as a guiding philosophy omits part of the very essence of what EID is all about. Indeed, it is this very taxonomy that provides the foundation for the discussions presented in Chapters 6 to 8 of this book, and, as the reader will see, it is the SRK taxonomy theory in particular that gives rise to the interesting questions surrounding the design of interfaces that use sensory modes other than that of vision.

This chapter has gone some way to argue why the SRK is an important part of the EID method; however, we would go further and argue that even the theory alone, without the WDA part of EID, can still provide significant guidance to system justification and design. In particular, and as will be seen in the coming chapters, the SRK theory can be used to justify, based on its description of human behaviour and cognition, certain design choices and presentation methods. As was described in Chapter 2, and mentioned in the introduction to this chapter, the remainder of this book does *not* therefore provide an example of the application of EID, or a test of the method as a whole, rather it offers an exploration of the potential for different types of information to support eco-driving behaviours at different levels of cognitive control.

In terms of the journey of the book as a whole, the current chapter has presented the first exploration of the design method that was, at the outset of this project, considered to be suitable for the design of in-vehicle information systems (see Chapter 2). The importance of the SRK taxonomy, and the first of EID's principles (i.e. to support SBB via the presentation of time-space signals, and allow the user to act directly on the display), provides the ongoing focus of the research presented herein; however, where this chapter has been devoted solely to the theory, the next chapter applies the theory to the eco-driving domain. To this end, the following chapter presents the analysis of expert eco-drivers' decision-making processes, in terms of the SRK taxonomy and the support of behaviour at lower levels of cognitive control, with the aim of building an understanding of the strategies used by those that are more experienced in fuel-efficient driving.

6 A Decision Ladder Analysis of Eco-driving
The First Step Toward Fuel-efficient Driving Behaviour

6.1 INTRODUCTION

The early stages of the research project described in this book began with the initial aim of guiding the design of in-vehicle driver support systems for vehicles with unconventional drivetrains (i.e. hybrid and electric vehicles). Drawing on the theory behind ecological interface design (EID), it was suggested that the method would be suitable for application in the design of in-vehicle systems for low-carbon vehicles. If one were to follow EID as described by its originators (see Chapter 5) one would first perform a work domain analysis (using the abstraction hierarchy, abstraction decomposition space or both). Indeed, at the start of this project, this was the intention; to model the work domain of the low-carbon vehicle in order to inform the design of an in-vehicle system that safely helps the driver maximise the utility of the potentially limited-range vehicle. As this research project progressed, however, it changed in two primary ways.

First, low-carbon vehicles no longer provide the sole focus of the research effort described herein. Supporting efficient use is particularly important in vehicles with range limitations (such that the user can get the most out of what range they have), and in vehicles with novel interaction characteristics (i.e. by fostering new habits through taking advantage of the novelty of interaction; see Chapter 2); however, its benefits can be realised in *all* road vehicles. The focus is on the specific behaviours that characterise fuel-efficient driving itself, and how to encourage them. The decision was therefore taken *not* to perform a full work domain analysis of the driving domain, but to concentrate on how to support particular behaviours in the vehicle.

This also relates to the second way in which this research changed focus. As the reader will discover, rather than attempt to apply the full EID method, attention has been paid only to its three core principles (particularly the first; to support skill-based behaviour), and to the SRK taxonomy (Rasmussen 1983). The previous chapter went some way to provide a general discussion of the method and its applications, and to argue for the benefits that the SRK taxonomy can bring to the interface design process. The current chapter continues this theme, with the aim of providing design guidance, couched in terms of the SRK taxonomy, on *how* information is to be displayed (see Chapter 5). Specifically, how do we encourage, at the lowest levels of cognitive control, eco-driving behaviour in the vehicle?

The necessary first step to this aim is to apply the theory discussed in Chapter 5 to an eco-driving context. To do so this chapter uses a representation of the SRK taxonomy, Rasmussen's decision ladders (introduced in the previous chapter; Rasmussen 1974), to model the activities that most affect fuel consumption in the road vehicle. The decision ladders, briefly introduced in the previous chapter, model activity in decision-making terms. According to their originator, Jens Rasmussen, they can be used 'to facilitate the matching of the formatting and encoding of data displays to the different modes of perception and processing used by human process controllers' (Rasmussen 1974, 26). Thus the goal of the current chapter is to develop models of specific driving activities (i.e. those that have the most significant effect on fuel economy) in order to inform the design of in-vehicle information to support those activities in the novice eco-driver (i.e. someone who has little experience in driving with the primary goal of fuel efficiency).

Before describing the process by which specific activities were identified, and how they were analysed, a brief explanation of decision ladders is offered (see Chapter 5 for a discussion of the SRK taxonomy).

6.2   DECISION LADDERS

As aforementioned, the decision ladder model is a task analysis framework that is used as an aid to design, representing activity in decision-making terms (Rasmussen 1974). Originally developed as reference frame for designers of human–machine systems, it provides a model of human data processing activity that draws 'heavily on upon engineering analogies' (Rasmussen 1974, 5). It does so by presenting the information used, options considered and decisions made at different stages of a particular activity. Though the model considers the human actor as 'a data processor through which input information received from the environment is connected to the output' (Rasmussen 1974, 5), it is also accepted that 'man [sic] is far more than a mechanistic data processor' (ibid.).

An example decision ladder is presented in Figure 6.1. In the figure, two different types of nodes can be seen; the rectangular boxes represent information processing activities, whereas the circles represent the resultant state of knowledge. The left portion of the diagram is concerned with an analysis of the situation and diagnosis of the current state of affairs, while the right side deals with the definition, planning and execution of an action. The top of the diagram represents the evaluation of options and the consideration of specific goals pertaining to the task at hand. Although sequentially arranged, the entry and exit points, and the sequence actually followed, will depend upon the nature of the task and the nature of the actor. For example, in some situations the top part of the diagram may be circulated more than once. In these instances the decision maker may have to consider the various options available to him or her, and what effect each of these options will have on the chosen goal of the activity. Furthermore, there may be multiple, conflicting goals present in the decision-making task, each of which will require consideration. The result of this stage may therefore present a trade-off between these goals, with the actor deciding upon a course of action that may be either optimal or satisfactory.

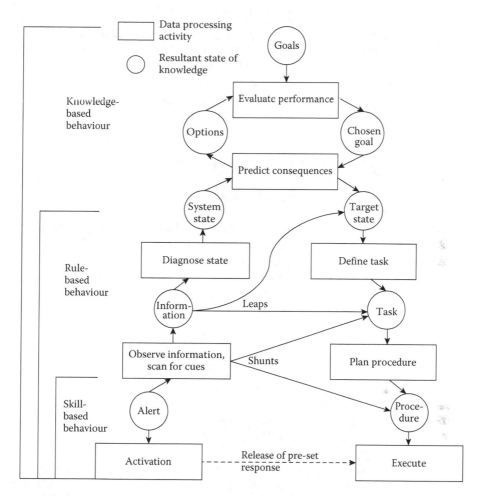

FIGURE 6.1 Decision ladder. (Adapted from Jenkins, D.P., et al., *Cognitive Work Analysis: Coping With Complexity*, Ashgate Publishing Limited, Farnham, England, 2009; Rasmussen, J., *The human data processor as a system component. Bits and pieces of a model. Riso-M-1722*, Roskilde, Denmark, 1974.) See also Figure 5.3, Chapter 5, this volume.

In familiar situations, and in experienced actors, the linear sequence depicted in the decision ladder is rarely followed; shortcuts are often taken. There are two types of shortcuts defined in the literature (Jenkins et al. 2009; Vicente 1999); shunts and leaps. Shunts connect data processing activities to non-sequential states of knowledge, whereas leaps connect two states of knowledge. The arrows in the centre of Figure 6.1 represent these shortcuts. For example, in certain situations the process of diagnosing the system state may lead directly to the knowledge that a set procedure is required; such a shortcut is an example of a shunt. It is a shortcut from a process to a state. An example of a leap would be the association of knowledge of the current system state with a knowledge of a task that needs to be performed in order to, for example, get the system back to normal system operations.

These shortcuts are often driven by rules and heuristics, learned through, for example, formal training and informal experience. They represent shortcuts from one state to another state.

Actors may also enter the decision ladder at different state of knowledge or information processing nodes; they do not necessarily have to enter at activation and exit at execute. For example, an actor may enter the decision ladder with an understanding of the current system state, or the structure of the task may be such that it is initiated by the knowledge of a goal state. From this they may infer, from past experience, the action required to achieve the given goal. Similarly, the activity may not necessarily flow from left to right, but can occur from right to left. For example, knowledge of the desired target state may lead an actor to observe for more information and cues to understand how this state may be achieved.

The aforementioned shortcuts represented in the decision ladder are indicative of rule-based behaviours; they represent instances in which familiar perceptual cues in the environment are associated with stored rules for action and intent. For example, if one is making the morning brew, an audible alert provided by a kettle leads to the knowledge that the water is boiling (system state); an understanding of the task (i.e. to pour the hot water from the kettle into the mug containing tea) immediately follows, with no need for the conscious consideration of options or goals. Skill-based behaviour, the fast, automatic response to stimuli in the environment, is represented on the decision ladder by the arrow connecting activation with execute. Here, upon activation of the decision-making process, a pre-set response is released, resulting in the execution of a particular activity. An example of this might be the falling of the mug from the kitchen table; upon the alert (i.e. one knocks the mug with an elbow and it begins to fall), the required action is immediately obvious and can said to be unconscious (i.e. to instinctively reach out to catch it).

As previously described, the full decision ladder, when annotated for a given decision-making process, will represent the way in which an actor analyses the situation, evaluates and selects goals, and plans and executes a task when using knowledge-based reasoning (i.e. follows the sequential path in its entirety), with all possible information inputs and options; this represents a prototypical model of activity (Jenkins et al. 2010). Rather than representing any one particular instance of an activity and the decisions therein (this would be a typical model of activity), the prototypical model aims to capture all possible elements that may affect the decision-making process (though not all will be used in any given situation).

6.3   IDENTIFICATION OF ACTIVITIES

In order to constrain and provide focus to the analysis, a number of specific driving activities were first identified; in particular, those that have the most significant effect on fuel economy. This process involved a review of both the academic literature and of more publicly available web resources (i.e. those not requiring subscriptions; free to access) pertaining to eco-driving.

Early research by Hooker (1988) suggested that it is the style of acceleration and the timing of gear change that have the largest effect on fuel use in the vehicle. More recent research concurs; according to Barkenbus (2010), eco-driving is characterised by (among other things) smooth acceleration, shifting up to the highest gear possible as early as possible (within the boundaries of safety), and anticipating the traffic flow and road layout ahead so as to avoid sudden starts and stops (i.e. to drive as smoothly as possible).

Anticipation for eco-driving is a concept that also features heavily across a multitude of publicly available internet resources, including specific eco-driving websites, (e.g. Ecowill 2015; Travelfootprint.org 2013), motoring organisations (e.g. The AA 2015), car manufacturers (e.g. Ford 2013; Renault 2013), local government (e.g. Devon County Council 2013), and from national and international non-governmental organisations (e.g. Energy Saving Trust 2015; United Nations 2013). The majority of these resources also provide information regarding the effect of vehicle maintenance, electrical system usage and weight management on fuel economy; however, as this research is only concerned with the types of driving styles and behaviours that characterise fuel-efficient use of the vehicle, that is, the driving task itself, these maintenance and peripheral use related considerations were not included in the current study.

From the sources listed above, two main classes of behaviour that have significant effects on fuel economy were identified; those relating to the brake and accelerator pedal, and those relating to the use of gears. Though the issue of gear choice is indeed significant in terms of fuel use, this class of behaviour was not considered in the current study for two reasons. First, this reduces the complexity and ensures focus, and second, the ultimate aim is to develop a system that is equally useful in automatic transmission vehicles, and electric and hybrid vehicles (where gear change advice is not applicable).

Given these criteria, and based on the literature reviewed, four specific activities were identified for modelling with decision ladders, namely, acceleration from a standstill, headway maintenance, deceleration with a full stop being more likely (e.g. for a red traffic light or stop sign), and deceleration with a full stop less likely (e.g. from a higher to a lower speed limit road section).

6.4 METHOD

An initial attempt to model the decision-making processes involved in the four activities described above was made using decision ladder templates (like that displayed in Figure 6.1). This was performed solely by this book's first author. To assess, validate and further specify these models, two different information gathering activities were performed; first, a focus group involving researchers at the University of Southampton's Transportation Research Group, and second, a series of five interviews with eco-driving experts. These will each be discussed in turn.

6.4.1 FOCUS GROUP

6.4.1.1 Participants

The resultant, first-iteration decision ladder models were discussed in a focus group with four researchers (including both authors of this book). Each participant possessed a working knowledge of human factors in road transport, however none of

TABLE 6.1

Focus Group Participant Information

Participant	Gender	Age	Years Driving	Years Involved in Road Transport Research
1	Male	53	37	20
2	Male	27	4	2
3	Female	28	11	6
4	Female	25	8	2

the members of the focus group was an expert in eco-driving specifically. The group served both to validate the choice of activities, and to discuss the initial models. Table 6.1 provides a summary of the four participants' relevant information.

6.4.1.2 Apparatus and Setting

The focus group was held in an office at the University of Southampton's Transportation Research Group. The four decision ladder models created in the first iteration were done so on A3 printouts of the decision ladder template; one template for each activity. These annotated templates were discussed in the focus group, with any additions or alterations being recorded on the diagrams themselves.

6.4.1.3 Procedure

Initial, general discussions focused on the four activities' relevance, and the extent to which they covered the behaviours of interest. Upon agreement on these activities, the four decision ladders were again discussed in more detail. The session was relatively informal in nature and lasted for approximately 2 hours. Each annotated model was introduced and explained by the first author, followed by a discussion of each annotated node on the decision ladder. Additional information that was not previously considered but deemed important by the group was added, and information that was already present was either agreed upon by all participants, edited such that agreement was reached, or it was removed.

6.4.2 EXPERT INTERVIEWS

The focus group served as an initial model validation stage; however, as aforementioned, the participants were not eco-driving subject matter experts. As such, five interviews with experienced eco-drivers were conducted in order to further validate the models.

6.4.2.1 Participants

Participants were initially sought from two eco-driving websites: ecomodder. com and hypermiler.co.uk. These websites provide a platform for those interested in both the technologies and behaviours associated with fuel-efficient driving. A request for participation was posted to the forums hosted on each website.

TABLE 6.2

Interviewee Information

Participant	Gender	Age	Years Driving	Years Eco-driving	Motivation	Primary Car Driven
1	Male	45	30	27	Financial and environmental	2003 Honda Civic Hybrid
2	Male	72	>50	30	Financial and 'as a game'	2007 KIA Ceed 1.6 diesel
3	Male	45	27	7	Environmental and through work	2004 Ford C-Max
4	Male	42	25	9	Environmental and through work	2005 Audi A3
5	Male	41	23	8	Financial and through work	2005 Skoda Fabia VRS

From this, two individuals made contact via email; one was a member of the forums on ecomodder.com, the other on hypermiler.co.uk. Three more participants were contacted through the ECOWILL project, details of which can be found from www.ecodrive.org. This European-wide project aims at providing information on eco-driving to the general public, as well as undertaking formal, academic research into various eco-driving aspects, including research involving driving instructors trained and experienced in teaching eco-driving techniques. The five individuals who responded were therefore all included as participants, representing a convenience sample. Ethical approval for the interviews was sought from and granted by the University of Southampton's Ethics and Research Governance committee. In all cases, participation was entirely voluntary, without any payment (monetary or otherwise). Relevant participant information is provided in Table 6.2.

6.4.2.2 Apparatus and Setting

Due to the geographically dispersed nature of the participants (one in the United States, one in Germany, one in Scotland, two in England), face-to-face interviews were not possible; hence four interviews were conducted using Skype™, with the other conducted over the telephone (as per this participant's preference). The participants were all in their own homes at the time of the interview. The interviewer was situated at the University of Southampton Transportation Research Group, in a quiet office, for all interviews.

6.4.2.3 Procedure

Each interview lasted approximately 1 hour, with the procedure for eliciting the information required to populate the models closely following that described in Jenkins et al. (2010). The process started by having the expert describe his higher-order goal (or goals) for the activity under analysis (bearing in mind the focus on fuel economy) and any potential constraints that may affect the attainment of those goals. Then followed a two-phase procedure; first the expert was asked to recall and 'walk-through' a typical situation, following which each stage of the model was supplemented with additional information regarding other possibilities for observation and action.

The interviewees were first introduced to the un-annotated decision ladder model and provided with a short explanation of its purpose, aims and intended use. They were invited to ask questions about the technique should they have any. They did not see the fully annotated models (i.e. those that resulted from the first iteration and the focus groups) until after each task was discussed. At this point the interviewee was emailed the fully annotated models that had resulted from the focus group.

To annotate the 'alert' stage the expert was first asked to identify the point at which he becomes aware of the need for action, and what in the environment influences his decision to act. In the second phase the expert was asked if there were any other possible cues or alerts that may influence his decision to act other than that already stated.

The 'information' stage was populated by asking the expert to list the sources of information he used to inform himself of the current state of the system. This was again supplemented during the second phase by asking him for any other potential sources of information that may be of use in similar situations. As with Jenkins et al. (2010), when the expert made reference to system states rather than stop him, a note was made and he was allowed to continue undisturbed. This was quite common, with each of the experts at some point describing the information used in terms of the resultant understanding of the system. As such, though the two-phase procedure for annotating the 'system state' node followed as closely as possible the procedure applied for the previous nodes, it was not always possible to separate the two discussions (i.e. on 'information' and 'system state'). The second pass did still serve to validate that which had been recorded by the interviewer.

The 'options' stage was annotated by asking the expert of the options he considered in a particular situation, and how these would affect the high-order goal for that particular activity. The number of options available is affected by the system state and as such the second pass (i.e. supplementing with general information rather than information regarding a specific example) revealed a number of other options that could potentially be available to the driver.

To annotate the 'chosen goal' the expert was asked about which constraints receive the highest priority. For eco-driving this was greatly affected by the state of the system; for example, all participants noted that weather and traffic conditions greatly affect the extent to which eco-driving activities are performed. As such, the two-phase approach was less useful here; rather each potential goal was discussed in turn. The 'target state' generally mirrors the option selection process; however, for the purposes of this analysis the fuel economy goal was accepted as having priority (assuming that safety constraints are satisfied; this will always have priority in a real, on-road situation). The expert was therefore asked to go through a particular situation, followed by consideration of other possibilities (i.e. a return to the two-phase process).

The final two stages, 'task' and 'procedure', were highly inter-related insofar as the procedure lists the steps necessary to perform the aforementioned task. The process for annotating these nodes therefore occurred in parallel, with the expert providing information on both of these steps for a specific example in the first phase, and the second phase involving him supplement both of these nodes where possible.

Throughout this process the decision ladder model in question (that which had resulted from the focus group) was edited and added to by the interviewer; the annotated model (i.e. one that included the annotations made during both the focus group and the interview) was then shown to the interviewee after the process for each activity described above was complete. They were asked whether they agreed with the model's annotations and asked if there was any additional information or decision-making processes relevant to each task step. If so, these were added. The interview process then moved on to the next activity, and the procedure repeated.

6.5 RESULTS

Following the discussions it became clear that 'deceleration with a full stop less likely' (e.g. from a higher to a lower speed limit road section) was too broad a category, insofar as the decision-making steps applied when approaching a road curvature were sufficiently different to the information used in other slowing events to warrant its own decision ladder. As such, two separate models were created; 'deceleration for road curvature' and the more generic, 'deceleration to lower speed'. Table 6.3 presents a summary of the five activities identified, alongside a brief description of each.

TABLE 6.3
Eco-driving Activities Selected for Analysis

Driving Activity	Description
Deceleration to lower speed (Figure 6.2)	For example, when approaching a higher speed limit to a lower one, or approaching a slower-moving lead vehicle. Early release of the accelerator pedal is advised in order to take advantage of the vehicle's momentum to carry it down to the required speed, again, to minimise use of the brake pedal.
Deceleration for road curvature (Figure 6.3)	When approaching a bend in the road the driver will often have to decrease their speed in order to safely negotiate the curvature. Releasing the accelerator early, taking advantage of the vehicle's momentum, is advised; hence minimising the use of the brake pedal is also a primary goal here.
Deceleration (full stop more likely) (Figure 6.4)	For example, when approaching a stop sign or traffic light at red. Early release of the accelerator pedal to take advantage of the vehicle's momentum to carry it to the stopping event is advised, to minimise use of the brake pedal.
Acceleration (Figure 6.5)	Either from a standstill, or from a lower speed to a higher speed. Though advice on fuel-efficient acceleration varies across information sources, there is a consensus that harsh, abrupt acceleration should be avoided.
Headway maintenance (Figure 6.6)	Though this does not have a direct effect on fuel economy, the indirect effect of maintaining a sufficient distance to the lead vehicle allows for early responses to upcoming events and affords the driver a better view of the road ahead (i.e. it is less blocked by the lead vehicle) therefore again supporting early responses to upcoming road events. This is also largely about minimising the need for brake pedal depression.

The five resultant decision ladder models are presented below in Figures 6.2 to 6.6, in the order they appear in Table 6.3. It is important at this point to clarify what is meant by the term 'coasting' in this chapter. Here it signifies the act of removing the foot from the accelerator pedal in order to make use of the car's momentum to carry it down the road *while still in gear*. It does not signify the act of putting the car into neutral, either through use of the gear stick or through full depression of the clutch pedal (the more common definition of coasting in the public media); this is often considered dangerous (due to a lack of vehicle control) and is generally not recommended (e.g. Gov.uk 2014). Furthermore, the act of coasting in gear has been shown to be more economical than coasting in neutral (Lin et al. 2011), though note that academic research is limited, and that specific fuel consumption rates while coasting in neutral or in gear are likely to depend on car age and type.

6.6   ANALYSIS

Figures 6.2 to 6.6 display the completed decision ladder models for each of the afore-mentioned activities. Shortcuts (i.e. shunts or leaps) are shown in the diagrams with either dashed or solid lines. Dashed lines indicate where the actor jumps from the left of the diagram to the right, reflecting where a particular activity or state of knowl-edge leads on to an activity further along the decision-making process. Solid lines are indicative of cyclical decision-making processes, that is, where an actor moves from performing some action (on the right of the diagram) back to an analysis of the situation (the left of the diagram). Each of the diagrams will now be discussed in turn.

6.6.1 Deceleration to Lower Speed

It can be seen from the left-hand side of Figure 6.2 that once the alert has been raised that there is upcoming need to slow down (i.e. a lower speed limit sign, or slower-moving traffic ahead has been seen by the driver), the driver scans for cues, both within and outside the vehicle, to build an understanding of the system state; this can also be thought of as developing an awareness of the situation. These informa-tion sources allow the driver to establish an understanding of the state of the system, which will in turn allow him/her to judge the distance required to coast down to the lower speed. Information from cues in the environment, in conjunction with the driver's previous experience on the road, will also allow the driver to infer the behav-iour of other road users; this will be particularly relevant in a situation where a driver is approaching slower-moving traffic ahead.

In the top part of the diagram the driver may cycle through the potential options for action, and consider the effect that the current system state will have on these pos-sibilities. For example, based on an understanding of the system state and experience of a particular vehicle's characteristics, the driver can estimate the effect of engine braking and different levels of hydraulic (i.e. traditional, brake pedal-initiated brak-ing) and regenerative braking (where this is applicable) on the state of the system as a whole. The effects of the current weather conditions on the driver's ability to perform

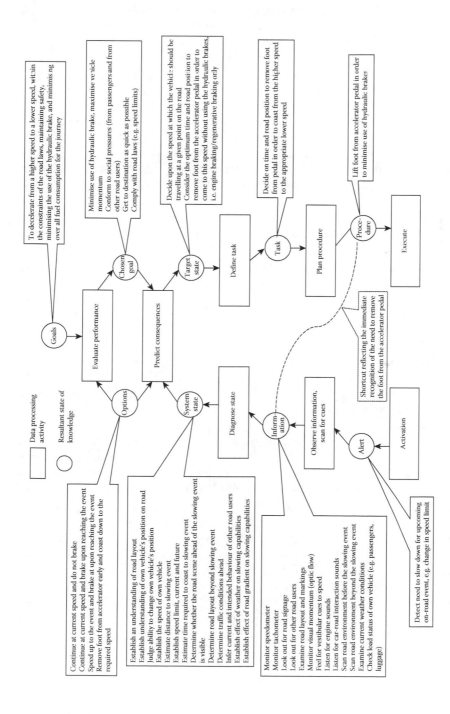

FIGURE 6.2 Decision ladder for deceleration to slower speed.

certain actions would be considered here, as well as their effect on fuel economy, legality, and safety of the different options available to the driver (i.e. continuing at current speed, braking late or coasting early). Furthermore, an assessment can be made regarding the effect on passengers, on other road users and on the impact a given option will have on arrival time at a particular destination.

As aforementioned, the overriding goal (in terms of this analysis) is to be able to decelerate, in the most fuel-efficient manner, down to a speed that is appropriate for the road ahead, for example in the case of slower-moving traffic ahead or a decrease in the speed limit. This is achieved by minimising the use of the hydraulic brake pedal, or conversely, maximising the use of the vehicle's momentum to carry it down to the speed required by the upcoming road situation. Three other potential goals have also been identified, two of which may be conflicting with the efficiency-related goal, namely arriving at the destination as quickly as possible, and conforming to social pressures.

In terms of time to arrival, one can imagine various situations in which speed may become the chosen goal, from the emergency (e.g. a pregnant woman, going into labour, being rushed to hospital by her partner) to the relatively mundane (e.g. rushing home from work in order to get back in time for the plumber's arrival).

With regard to social pressure, this can come from both within and outside the vehicle. For those pressures coming from within the vehicle, one can imagine, for example, a situation in which a young driver succumbs to peer pressure to drive more aggressively (an established finding, particularly for young men; e.g. Conner et al. 2003). Pressures coming from outside the vehicle relate to the behaviour of other road users, for example other drivers' use of their horns, or the act of driving very close to the rear of the driver's vehicle to encourage the driver to go faster (see, e.g. Åberg et al. (1997) for a discussion on the effect of the social environment on driver behaviour and perceptions).

The remaining possible goal, to comply with road laws, for example, speed limits, should be present for any activity taking place on a public road, however it has been specifically mentioned here as it is a guiding principle in the timing of decelerations for efficiency; the point at which the driver must be driving at a particular speed is governed by the speed limit signs (given traffic conditions will allow it). Hence it is the legal constraints of the road that place on the driver the need to slow down. Though the speed limit may be lower than the speed at which fuel use is optimal (hence in conflict with a pure fuel-efficiency goal), this goal is not in conflict with the goal of decelerating in the most fuel-efficient way. It simply informs the driver of when he or she must initiate an action.

Moving down the right-hand side of the diagram, the target state can be understood in terms of the use of the accelerator pedal, or more specifically, the time and road position (dependent on current speed) at which the foot should be removed from the accelerator pedal in order to coast, from the current speed, down to the speed necessitated by the upcoming road situation. This requires both an understanding of the speed which the vehicle must decelerate to (either read from a speed limit sign, or judged based on the characteristics of the upcoming traffic situation), and the deceleration characteristics of the vehicle when using only engine braking (i.e. without the use of the hydraulic brake). This knowledge of the target state necessarily leads on to an understanding of the task, that is, when to remove the foot from the accelerator pedal, and the procedure, that is, remove the foot and minimise hydraulic brake use.

It came out in the interviews that much of this process is automatic, or at least approaching automaticity, in the experienced eco-driver. Rather than discuss the full decision-making process, the interviewees would describe the process of scanning the environment for information, then immediately go on to explain their procedure, that is, to remove their foot from the accelerator as early as possible. This is reflected by the shortcut from 'information' to 'procedure'.

6.6.2 DECELERATION FOR ROAD CURVATURE

For this activity (Figure 6.3) the overriding goal is 'to decelerate from a higher speed to a lower speed in order to negotiate a road curvature while maintaining safety and minimising overall fuel consumption for the journey'.

This diagram shares many characteristics with that for deceleration to a lower speed, insofar as an event is detected, information is sought both from within and outside the vehicle, and the state of the road environment and its users are understood. The most significant difference here is the need to judge the angle of the upcoming corner and, relatedly, to judge the speed which the driver must decelerate to in order to negotiate the corner safely (should deceleration be required). If possible, the driver will also scan the road environment beyond the corner for information, in terms of both road layout and traffic conditions. The weather (and, relatedly, road surface) conditions may also influence the driver's understanding of the vehicle's cornering capabilities (e.g. icy conditions would likely necessitate a more cautious cornering manoeuvre).

The top part of the diagram also resembles that of Figure 6.2; however, a notable difference is the need to determine the effect of the weather on the car's cornering abilities (largely related to grip) and the effect of approach speed on the driver's ability to achieve the desired cornering line, in terms of maximising the view of the road ahead, ensuring safety and maximising fuel efficiency. One could imagine a situation where heavy rain, for example, would necessitate a more cautious cornering strategy. Though this would incur greater fuel use (maintaining momentum will always offer the most fuel-efficient strategy), in these situations safety would be of primary concern.

The goal of maximising the view of the road ahead is also included here; this came out of the interviews as an important tactic for ensuring the driver's ability to anticipate the road and traffic situation beyond the corner. Though not explicitly described as such in the interviews, this relates to considering the risk involved in having an obscured forward view of the road environment. This arises from a combination of information sources, namely the understanding of the road layout and traffic conditions beyond the event. The shortcut from 'execute' to 'information' reflects the need to cycle through the diagram as one progresses round the corner (i.e. to continue to scan the road ahead). Should a driver find that they have to come to a full stop, they would transition from the cornering task to the alert stage of the 'deceleration for full stop more likely' task. Though the driving activities have been

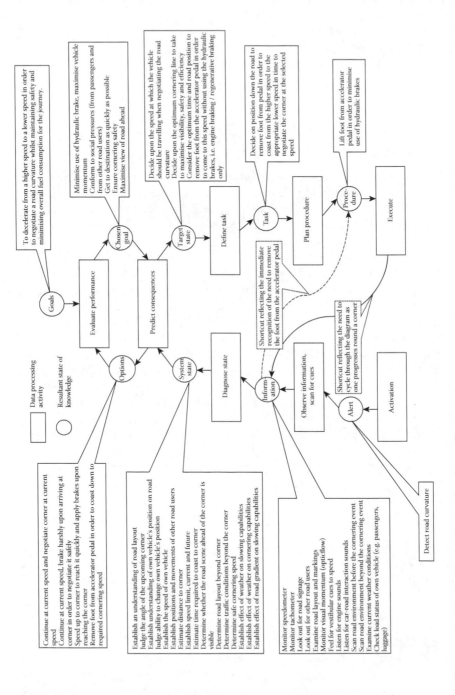

FIGURE 6.3 Decision ladder for deceleration for road curvature.

separated for the purposes of this analysis, in reality the various driving activities a driver will perform on a given journey will link and overlap.

In terms of tasks and procedures, as with the decision ladder for deceleration to a slower speed, the target state of understanding the optimum time and road position at which to remove the foot from the accelerator pedal (to minimise the need for hydraulic braking) leads on to an understanding of the task, that is, to remove the foot from the pedal at the optimum time. This then necessarily leads on to the procedure, that is, to decelerate from a higher speed to a lower speed without the use of the hydraulic brakes.

As with the process for 'deceleration to lower speed', only upon probing did the interviewees describe more than simply reacting to environmental stimuli with a pre-set response, that is, to remove the foot from the accelerator pedal. Hence the shortcut in Figure 6.3 is in the same position as that in Figure 6.2.

6.6.3 DECELERATION FOR FULL STOP MORE LIKELY

The primary, overarching goal for the activity of decelerating for an event likely to require a full stop (Figure 6.4) is 'to safely come to a stop (or slow down sufficiently early that a complete stop is not required) for a red traffic light or stop sign (or other event requiring a complete stop), with or without queuing vehicles, minimising the use of the hydraulic brake pedal'.

Again, the alert stage is characterised by the recognition of the upcoming road event, and the cues in the environment used by the driver to develop an understanding of the current system state are the same as those for the previous two decision ladders for deceleration. There are, however, different aspects of the system state that are important in terms of approaching an event that is likely to require a full stop; these reflect the considerations made that are unique to traffic light junctions and crossings, and when approaching a queue of static traffic ahead. This knowledge of the system's state will go on to inform the driver of the best or most suitable course of action. For example, in a familiar situation a driver may know approximately how long a light remains red; when they approach the light at red, they may use this information to guide their approach speed such that they reach the light at the point at which it changes to green, thereby minimising brake pedal usage and maximising efficiency.

At the top of the diagram the picture is very similar to that for decelerating to a slower speed; the main difference here is the consideration of the effect that a certain course of action will have on the likelihood that a full stop can be avoided. This is reflected in the goals section on the top right of the diagram; here there is an additional goal of 'maximise likelihood of not requiring a full stop'. Target state annotations follow suit; to act early enough that a full stop is not required is one target state, as is the use of momentum to carry the vehicle to the event, with the arrival at the stopping event being achieved without use of the hydraulic brakes. As with previous decision ladders, the task involves planning for the moment at which to remove the foot from the accelerator pedal, with the procedure being enacted at the correct moment in time and space; once again, in the experienced eco-driver this occurs without conscious consideration of all the options and goal states, hence the shortcut from 'information' to 'procedure'.

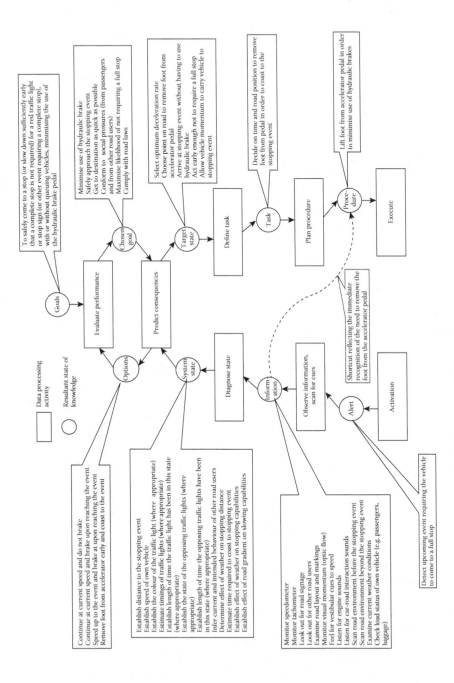

FIGURE 6.4 Decision ladder for deceleration for full stop more likely.

6.6.4 ACCELERATION

In the decision ladder for acceleration (Figure 6.5), it can be seen that the primary goal is 'to accelerate from a standstill, with the highest possible degree of fuel efficiency, within the boundaries of safety'. In the alert stage, the driver recognises both the need and the ability to move off from a standstill. This could be, for example, from a traffic light as it turns green, or from a road junction at which the vehicle is at a standstill.

Although the start of this decision-making process is defined as the vehicle being at a standstill (at the alert node), it is characteristically cyclical in nature, insofar as once the driver has started to move off (i.e. initial execution of the task) he or she will return to gathering information from the environment (particularly that which is indicative of acceleration, e.g. vestibular cues, visual momentum and car–road interaction noises) to help guide behaviour throughout the entirety of the acceleration activity (i.e. until target speed has been reached). This is indicated by the solid line shortcut in Figure 6.5, from 'execute' to 'information'.

When the target speed has been reached the driver will maintain said speed. Maintenance of speed would constitute a separate decision ladder diagram, one that the 'acceleration' decision ladder would feed into. It has not been included here for the sake of focus and brevity; however, one might expect an experienced individual, once they have cycled through the 'acceleration' decision ladder up to their target speed, to move from the 'execute' node on the acceleration decision ladder to the 'information' node on a 'speed maintenance' decision ladder.

Though the speed limit often guides target speed, it may not always; for example, in conditions of slower-moving traffic or certain weather conditions (e.g. icy roads), the posted speed limit may not present a safe or desirable target speed. Hence 'establish target speed' has been included as a distinct aspect of the knowledge of system state.

In the upper part of the diagram the possible actions need to be considered in terms of their effect on other road users, fuel use, legality and safety. There is, however, a potentially ambiguous issue to consider, namely the effect on fuel economy of accelerating more positively (i.e. quickly and harshly) into a smaller gap (when exiting a junction to merge with moving traffic), or waiting for a larger gap into which to accelerate more smoothly (which will itself be more efficient, but will incur more idling time). Furthermore, it is not only fuel efficiency that could be affected here; impacts may be seen on all the goals presented on the upper left part of the decision ladder, namely safety, journey time and the effect on others, that is, passengers and other road users.

As is the case in the previous decision ladders, information relating to the task node is reflected in the target state node, which in turn dictates the necessary procedure. This is described by the need to depress the pedal with the amount of force and rate of depression necessary to reach the target speed quickly and within the boundaries of safety, legality and efficiency. The reader will have noticed the two shortcuts in this decision ladder that link both 'alert' and 'information' with 'procedure'. These represent the interviewees' tendency to once again go straight from describing the need for action to describing the required procedure. The shortcut

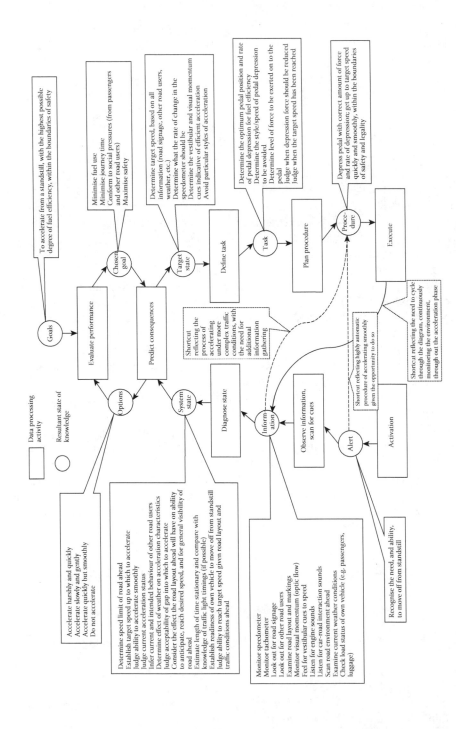

FIGURE 6.5 Decision ladder for acceleration.

from 'alert' reflects the automatic, over-learned procedure of smooth acceleration. The action is triggered by, for example, a traffic light change or the moving-off of the car in front; this stimulus–response pairing is an example of skill-based behaviour. When the road environment is more complex, for example there are more road users around or the weather conditions necessitate greater care and attention, the shortcut from 'information' is more apt. In these situations it is the combination of cues that dictate the required procedure.

6.6.5 HEADWAY MAINTENANCE

The overriding goal for the activity of headway maintenance (Figure 6.6) is to 'minimise the use of the hydraulic brakes through maintaining a safe headway to the lead vehicle, thus allowing early action to changes in the behaviour of the lead vehicle and other road users through both increased distance to act and a less obscured view of the road environment'.

Once a vehicle has been detected ahead (the alert stage) the driver again scans for information from within and outside the vehicle. The difference here is the focus on the lead vehicle and the road ahead of the lead vehicle. For this activity the primary concern is, of course, the behaviour of the lead vehicle; the driver must build an understanding of the level of safety of the separation between the lead vehicle and their own, and the effect this time and distance will have on both the ability to view the road environment ahead and on the ability to act early to system disturbances.

The goals for this activity are concerned with maintaining a distance to the vehicle that is conducive to early action, allowing for an unobstructed (or at least *less* obstructed) view of the road ahead of the lead vehicle. Depending on the context, the 'conform to social pressure' goal may have more influence over the decision-making process for this activity. For example, should there be a lot of traffic on the roads, the driver may be less willing to leave a large gap, as not only may others view this behaviour negatively (and potentially make their feelings known to the driver), in a multi-lane road this gap may be filled by another vehicle (thereby reducing separation distance to an undesirable level, thus necessitating additional deceleration). This would represent an 'alert' that would require the driver to perform the decision-making process anew.

Progressing down the right leg of the diagram it can be seen that the target state is to maintain a safe distance to the lead vehicle, achieved through applying the correct amount of force to the accelerator pedal. The task description also requires from the driver an understanding that they must remove the foot from the accelerator pedal at the opportune moment should the lead vehicle, or another vehicle in the road ahead, reduce its speed. This task plan is then translated into the procedure, namely to schedule the time at which to remove the foot from the accelerator pedal and the exertion of the appropriate level of force.

The solid-lined shortcut from 'execute' to 'information' reflects the cyclical nature of the task, insofar as the driver must continue to monitor information sources in order to maintain the required headway separation, performing actions as necessary (i.e. the manipulation of the accelerator pedal and, potentially, the brake pedal).

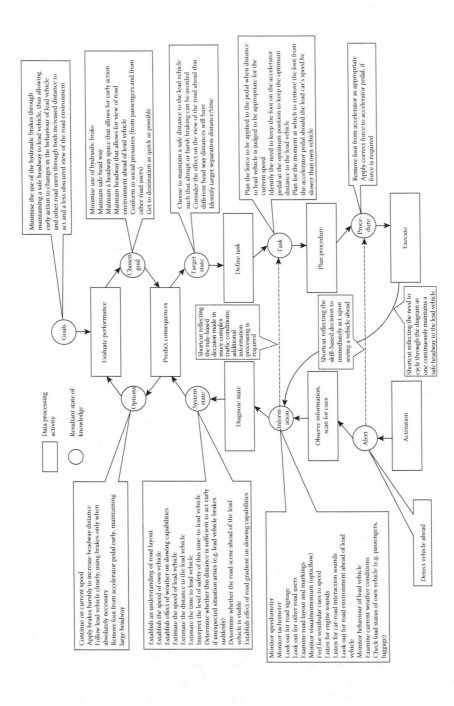

FIGURE 6.6 Decision ladder for headway maintenance.

The two shortcuts going from left to right once again reflect the interviewees' tendency to talk about the procedure and task without first discussing state diagnosis or option and goal considerations. In most instances the interviewees simply stated that they applied the 'correct' amount of force and maintained a 'correct' distance to the lead vehicle, manipulating the accelerator pedal as necessary; stimuli in the environment immediately suggest a response. This skill-based behaviour is indicated by the shortcut from 'alert' to 'procedure'. The shortcut from 'information' to 'task' is indicative of behaviour in a more complex road environment, for example in heavy traffic or adverse weather conditions; here, the driver must pay more attention to information in the environment, with a number of cues suggesting to the driver a suitable course of action.

6.7 IMPLICATIONS FOR DESIGN

As described in the introductory sections of this chapter, the way in which an individual progresses from the alert stage to the execute stage will depend on a number of factors, from the characteristics of the driver (e.g. novice or expert) to the information available at a specific time and location (e.g. signage may differ, visibility may be different depending on time of day or weather). The question of importance here is how can we design a driver support system in such a way as to support different paths through the model (i.e. to follow the shortcuts presented in Figures 6.2 to 6.6)? Given the right information presentation method, it may be possible to support skill- and rule-based eco-driving behaviours even in the novice driver. The primary aim is, therefore, to transform a cognitive task into a perceptual task.

As aforementioned, in almost all instances the expert eco-driver would describe his specific action (i.e. procedure) immediately after his description of the detection of an upcoming event (hence the shortcuts presented in the decision ladder figures). In the situations requiring deceleration, for example, this was to remove the foot from the accelerator pedal, even if the event were over 500 metres away (reported by interviewees 2 and 3).

It is this automatic, early response to the alert that characterises expert eco-driving. For example, Figures 6.2 to 6.4 describe how it is the timing of the foot's removal from the pedal that is critical. It may therefore be possible to provide a stimulus to the driver to encourage them to remove their foot from the pedal in the same way an expert eco-driver would do automatically (or at least in a manner approaching automaticity). A system that provides such information can be said to support shortcuts through the decision ladder, thereby potentially eliciting behaviours that would otherwise only be produced by those with eco-driving expertise. Given that information in the environment will also be present (i.e. that which the expert uses to guide behaviour) the system as a whole will still support behaviour at the rule- and knowledge-based levels, therefore offering support for learning through association (i.e. associating information system cues with environmental cues).

Generally speaking, it is possible to frame this idea as follows; shortcut X in Figure Y can be supported by design Z. Take Figure 6.2, deceleration to lower speed; here, the shortcut from 'information' to 'procedure' could be supported by the presentation of a stimulus, provided at the optimum moment for fuel efficiency, that suggests to the driver that they remove their foot from the accelerator pedal. The same system would also function in the situations described in Figures 6.3 and 6.4; the shortcuts through the ladders, currently taken by expert eco-drivers through their ability to amalgamate multiple sources of information into a cue for action, could be supported by well-timed information presentation.

This kind of system would therefore have potential to support skill-based eco-driving behaviour in the novice eco-driver, as it would do so through supporting interaction via time-space signals, a necessary means for encouraging skill-based behaviour (Rasmussen 1983; Vicente and Rasmussen 1992). In the majority of situations the timing of the presentation of such information could be calculated using information that is already available from car radar systems and satellite navigation information. Indeed, in Muñoz-Organero and Magaña (2013) a retroactively fitted information system is described that does just this. The system detects upcoming traffic lights and provides the driver with advice concerning the optimal deceleration patterns required to efficiently come to a stop; considering the analysis presented in this chapter, this system can be justified using decision ladders and the associated SRK theoretical framework. The system, in effect, takes a collection of cues from the environment and converts them into one signal for action.

As has been described above for the deceleration events, the shortcuts in Figures 6.5 and 6.6 can also be used as a starting point for the design of an acceleration and headway maintenance support system. To encourage these behaviours in an efficient way a similar system to that described above can be envisaged. Again, taking advantage of the ever-increasing sophistication of radar, satellite navigation and vehicle-to-vehicle and vehicle-to-infrastructure technology, it may be possible to display to the driver the suggested maximum amount of accelerator pedal depression for a given scenario. For example, it has been suggested that depression of the accelerator pedal beyond 50% of its travel represents an inefficient acceleration strategy (Birrell et al. 2013). Though this may be an oversimplification (different engine and drive-train technologies will have their own efficiency patterns), and at times may not be the optimum strategy (e.g. safety constraints may necessitate harsher acceleration), the process applied in Birrell et al.'s study (2013) of vibrating the accelerator pedal when acceleration levels exceed a given threshold represents a method of supporting a shortcut through the decision ladder; where the expert eco-driver amalgamates information from different sources to guide behaviour, the novice is supported in performing the same behaviour by a simple, in-vehicle, time-space signal.

The same applies for headway maintenance; the driver could be provided with a signal to indicate when they are undesirably close to the lead vehicle, based on various factors including, for example, current speed of both vehicles, road traffic laws and weather conditions. Mulder and colleagues describe such a system in their work (Mulder et al. 2008, 2011). In this research, car-following behaviours (i.e. headway maintenance behaviours) are actively supported by haptic feedback presented through the accelerator pedal (in this case, stiffness or force feedback). Though the focus was on safe rather than fuel-efficient driving, the two styles share many characteristics (e.g. Young et al. 2011). Note that both of these potential systems only provide to the driver an indication of when they are using the accelerator pedal excessively; they would not inform the driver if they were depressing the pedal *insufficiently*. Though this is not considered to be especially problematic (the more significant problems for safety and fuel efficiency come from the excessive use of accelerator pedal rather than from over conservative driving behaviours), one system that does provide such information to the driver is that described by Seppelt and Lee (2007). In their paper Seppelt and Lee describe a visual representation of an adaptive cruise control (ACC) system that displays the behaviour of the system in terms of time headway, time to collision and range rate. Though not explicitly aiming at supporting headway maintenance behaviours (rather it was focused on ensuring awareness of ACC system functioning) it was shown to support faster and more consistent braking responses when the system required them. Interestingly, this system was designed using EID; Rasmussen and Vicente (1989); Vicente and Rasmussen (1992); see Chapter 5.

6.7.1 SUPPORTING SKILL-BASED BEHAVIOUR WITH HAPTIC FEEDBACK

The SRK taxonomy becomes increasingly important as we begin to see the use of different sensory modes in system interfaces, particularly the haptic mode (relating to the sense of touch) as seen in Birrell et al.'s study of vibrotactile feedback for efficient acceleration. As described in Chapters 2 and 5, the EID approach to design (drawing heavily from the SRK philosophy) is based on the tenets of Gibsonian ecological psychology (Gibson 1979). The central concept is that when an interface is designed in keeping with the tenets of EID, the environmental constraints acting on the system are represented in such a way that direct perception is possible; this removes the requirement on the user to create and maintain indirect mental representations of the system and the external reality. To put this in other words, EID aims to represent the environment or system in a way that matches human perception.

There has, however, been a largely visual focus across the extant EID literature. This presents an interesting issue; as Olsheski (2012) pointed out, daily interaction within the environment is almost never unimodal, hence if the technique is to truly represent an ecological approach to interface design then we must move away from this unimodal focus. Sanderson et al. (2000) also make this point, arguing that we, as humans, have evolved to process information from many different modes. It is therefore surprising that EID, aimed at representing the world in a way that matches human perception, has thus far largely neglected audition (though see Sanderson 2006; Sanderson et al. 2000, 2009; Sanderson and Watson 2005; Watson et al. 2000a, 2000b) and, with the exception of work by Lee et al. (2004), has almost completely neglected the haptic channel.

Lee et al.'s (2004) work is of particular significance when considering the contribution the SRK framework could make to the design of haptic interfaces. In their paper, they suggest the work domain analysis part of EID provides *what* to display while the SRK framework provides guidance on *how* to present that information (as was discussed in the previous chapter). That EID aims to display the boundaries or constraints acting upon a system in a way that is directly perceptible (thus allowing for perceptual-motor driven control) is also particularly interesting when thinking about haptic information in the private road vehicle; in this context the concept can be expressed as allowing the driver to 'feel' the 'field of safe travel' (Gibson and Crooks 1938).

The field of safe travel concept is also described in Birrell and Young's work (Birrell and Young 2011; Young and Birrell 2012) in relation to a visual, EID inspired in-car interface; however, in our judgement it is in the haptic display of such constraints that lies the more interesting relation to the SRK framework. Moreover, the idea of feeling system boundaries applies to eco-driving (as discussed in Chapter 2) just as it does to safe driving. Rather than presenting the boundaries of the field of *safe* travel, one can present to the driver the boundaries of the field of *efficient* travel.

6.7.2 HAPTIC INFORMATION IN VEHICLES

Given Birrell et al.'s work with vibrotactile feedback and Mulder et al.'s work on stiffness and force feedback, both presented from the accelerator pedal, parallels can be drawn with the theory behind direct manipulation interfaces (DMI; e.g. Hutchins et al. 1986). This approach emphasises the need to represent objects of interest and to allow the users to act directly on what they can see in a display; it provides both an 'attempt to display the domain objects of interest and allow the operator to act directly on those objects' (Rasmussen and Vicente 1989, 527) and allows the operator 'to rely on the perceptual cues provided by the interface to control the system' (525, ibid.). Note that these quotes come not from DMI proponents, but from the originator of the SRK taxonomy and the creators of EID i.e. Kim Vicente and Jens Rasmussen.

The theory behind both EID and the DMI approach argues for the benefits of taking advantage of the human sensorimotor system, that is, to encourage behaviour at the skill-based level, and both argue for the combination of action and control surfaces (i.e. the site from which information is retrieved should be the same site on to which control actions are performed). When one considers that the tasks described in this chapter are all related to the use of the accelerator pedal, the possibility of providing tactile or haptic feedback through this very site becomes one that satisfies both the tenets of EID (and, in turn, the SRK taxonomy) and those of the DMI approach. This type of feedback should therefore, if one follows the DMI and SRK theoretical arguments, support skill-based eco-driving behaviour.

Furthermore, an argument can be made about the way different information presentation modes activate certain levels of cognitive control in the actor (in this instance the driver). Rasmussen (1983) describes how information can be interpreted by workers as either symbols, signs or signals; symbols activate knowledge-based reasoning, signs activate rule-based reasoning and signals activate skill-based reasoning. In Naikar (2006) a cooking analogy is used to describe these concepts; when pouring milk into a jug, a signal to indicate when to stop pouring would be the closing distance between the milk level and a measurement marker on the jug itself. A sign for this might come from an electronic scale that provides an auditory tone when the correct weight is reached, and a symbol may be stored as part of a mental model regarding the calculation of the amount of milk required for a certain recipe compared to the amount of milk in the carton.

In terms of the driving scenario, a physical stimulus presented through the accelerator pedal (indicating that the pedal should be depressed no further) would represent a signal insofar as it would form a physical barrier to further depression (not unsurpassable, but a noticeable barrier as much as the line in the milk jug is). It is a perceptual indicator of a correct or desired action. Moreover, Naikar's (2006) use of an auditory tone as an example of a sign is interesting when discussing feedback in the vehicle. This type of arbitrary stimulus has the potential to be used in the vehicle for the same purpose as the accelerator-based haptic feedback, namely to indicate to the driver when to remove their foot from the accelerator pedal. This may do so through supporting rule-based eco-driving behaviours.

Displaying the field of safe travel to the driver through haptic feedback in the vehicle offers a means for supporting a direct, analogical link between the environment and the interface. Lee et al. (2004) argue that in-vehicle haptic feedback is uniquely placed to allow for the combination control and observation surfaces, therefore supporting skill-based behaviour. As aforementioned, this allows for time/space signals to guide effortless responses to system changes; however, as described above, in Vicente and Rasmussen's work (Rasmussen and Vicente 1989; Vicente and Rasmussen 1992) only visual interfaces were discussed, with mouse and tracker-ball interaction methods given as examples of acting directly on the interface. The benefit in vehicles is that haptic information presents a means for not only combining action and observation surfaces, but action and observation sensory modes as well. That is to say, a haptic task (e.g. depressing an accelerator pedal or moving a steering wheel) can be supported with haptic feedback supplied at the same location. By coupling the vibrotactile or haptic information with the device used to act on that information strong mental models can be developed and supported; there is spatial contiguity (Mayer 2001; Seaborn and Antle 2011).

A number of other researchers in the driving domain have discussed this possibility, though not with the SRK taxonomy or EID in mind. Forsyth and MacLean (2006) investigated joystick-based haptic feedback for path following; Steele and Gillespie assessed steering wheel-based haptic navigation information (Steele and Gillespie 2001); and Hajek et al. (2011) studied the addition of force feedback to the accelerator pedal to indicate optimal braking points for safety and efficiency. Other instances include the aforementioned work by Mulder and colleagues (Mulder et al. 2008, 2011) on the addition of force feedback in the accelerator pedal to enhance car-following safety behaviours, as well as research by Várhelyi and colleagues (Adell and Várhelyi 2008; Adell et al. 2008; Várhelyi et al. 2004) on accelerator-based haptic feedback for speed compliance, and work by Van Winsum (1999; cited in Van Erp and van Veen (2001)) showing that shorter reaction times are supported by accelerator pedal-based haptic feedback that is indicated when the speed limit is exceeded.

Similar arguments were also forwarded by Tijerina (1995), who argued that haptic information should correspond to the required action. For example, if steering is required then vibrotactile information should be presented through the steering wheel, or if deceleration is required, force feedback should be added to the accelerator pedal (Tijerina 1995). Such arguments find similarities with the SRK and EID philosophy. Van Erp and van Veen (2001, 2004) also made this point, stating that though current controls and displays in a vehicle are often incongruous, tactile displays could circumvent this issue. They state that with tactile stimuli the cognitive resource requirements to extract meaning from the stimulus would be minimal. In this way the stimulus could be said to be in line with ecological principles;

it is immediately perceptible (thus allowing control at the rule- or skill-based level), without requiring the development of internal mental representations of the system and environment.

Van Erp and van Veen (2001) also mention the possibility of joystick control in vehicles, already a reality in some vehicles modified for use by handicapped individuals. In this scenario, fingers themselves could be provided with haptic information (as could the palm); this could be beneficial as not only are fingers very sensitive to tactile information, they represent precisely the site of control. This offers the possibility of a stimulus–response type system, an identifying characteristic of skill-based behaviour. It is important to note, however, that naturally occurring (i.e. not specifically added) haptic feedback is already present in the steering wheel, alerting drivers to vehicle behaviour and road conditions. It is therefore imperative that additional haptic information does not interfere with, mask or become masked by this already present, natural information.

Though Lee et al. (2004) offer a theoretical discussion of the applicability of EID to the design of in-vehicle haptic displays, they do not discuss in detail any specific interface forms or feedback sites; it is a general discussion of the ability of haptic feedback to support skill- rule- and knowledge-based behaviour. Moreover, while there are a number of discussions of accelerator-based haptic feedback in the literature, to our knowledge none of them also discuss EID or the SRK framework. This is despite the clear similarities between the discussions found in haptic accelerator research and the theoretical discussions arising from the SRK taxonomy and from EID. For example, Mulder et al. (2008; see also Mulder et al. 2011) describe the motivation for their research into accelerator pedal-based haptic feedback for car-following support in terms of allowing the driver to 'virtually touch their environment through the haptic interface' (1711). They go on to suggest: 'In the haptic gas-pedal design for car-following support, haptic information of the safe-field-of-travel boundaries provides them, first of all, with a complementary channel, besides vision, to determine these boundaries. Second, continuous haptic presentation of the boundaries enables continuous haptic perception of these boundaries. Third, by presenting the haptic information through the gas pedal, a direct connection is created between stimulus and response, that is, longitudinal control information is presented through the longitudinal control channel in such a way that the stimulus is compatible with the required response' (Mulder et al. 2008, 1711).

The references to system boundaries and to the combination of observation and action surfaces clearly resonate with the SRK-guided EID philosophy. Such theoretical similarities lead to some potentially interesting questions; for example, does information presented to different sensory modes support different levels of cognitive control? In other words, for a manual task (such as depressing an accelerator pedal) is it inherently more supportive of skill-based behaviour to offer information through the manual, that is, haptic, sensory channel? While this coupling of action and observation sensory modes may only be appropriate for certain tasks and in certain domains, where it is possible to do so (e.g. in the driving domain) significant benefits could well be realised if this theoretical connection is valid.

This kind of theoretical argument can be used to explain the results of many of the studies referenced above (Hajek et al. 2011; Mulder et al. 2008, 2011).

Moreover, additional support for the benefits of providing haptic feedback at the site of control (the foot pedal) can also be found in Birrell et al. (2013), and Janssen and Nilsson (1993); in Birrell et al. (2013) more economical use of the accelerator pedal was supported by haptic feedback, and in Janssen and Nilsson decreases in dangerous headway maintenance behaviours (largely dependent on accelerator pedal usage) were facilitated to the greatest extent by accelerator pedal-based haptic feedback (Janssen and Nilsson 1993).

Although a detailed discussion of the matter is not within the scope of this book, it is important to at least recognise work carried out in the neuropsychology domain pertaining to intersensory facilitation. The interested reader is referred to Ho and Spence (2008) for a discussion; however, here it is sufficient to state that a significant body of research points to the ability of haptic feedback to draw the attention of a user to visual events in the environment (e.g. Ho et al. 2006; Ngo and Spence 2010). This is important for a discussion on supporting behaviour at different levels of cognitive control in the vehicle; though a haptic foot pedal may well support skill-based behaviour, there remains the requirement to support all levels of behavioural control, from skill-, to rule- and knowledge-based behaviour. As mentioned above, in the road vehicle the visual scene outside the car offers a display of sorts to the driver. This supports rule- and knowledge-based behaviour insofar as information in the external environment can be viewed and deliberated upon; if a haptic display can draw attention to events in the external environment as well as immediately supporting behaviour then the overall in-vehicle interface can be said to be in line with the principles of EID.

The use of accelerator pedal-based haptic feedback in the vehicle can also be justified with EID in terms of its ability to support learning and skill acquisition. As has already been discussed, an interface in keeping with EID principles should support skill acquisition; through the aggregation of individual movements or actions into more complex routines. Rasmussen and Vicente (1989) have argued that the user must be able to experiment in order to optimise skill; for this to safely occur the limits of acceptable performance should be made visible, with observable and reversible effects. In terms of a haptic pedal the user would be able to depress, and lift his or her foot from the pedal, feeling where the added force feedback or vibrational alert is presented. This type of information therefore guides dynamic interaction at the sensorimotor level. Moreover, cues provided by the accelerator pedal can be integrated with visual cues in the driving scene; once this chunking of information has occurred it is not only the information from the pedal that informs the driver of the optimum behaviour, but the cues in the visual scene as well.

6.8 CONCLUSIONS

In terms of the journey of this book as a whole, the analyses described in this chapter represent the application of theory, described in Chapter 5, to the eco-driving domain, a feature of Chapter 2 and the focus of Chapters 3 and 4. In one respect this chapter is not so different from Chapter 4 (in which the verbal reports of 19 drivers were analysed), at least in terms of its motivations; both aimed at learning from those that perform eco-driving behaviours to a greater extent in order to support those that perform them to a lesser extent. The two approaches have been different insofar as this chapter has focused purely on 'expert' eco-drivers, and has taken a more theory-driven viewpoint. In particular, how can we support skill-based behaviour in the novice eco-driver, considering the theoretical arguments inherent to the SRK taxonomy and, relatedly, EID?

It is clear from the extant literature that haptic feedback presented through the accelerator pedal can successfully encourage certain behaviours (e.g. Birrell et al. 2013; Hajek et al. 2011; Jamson et al. 2013; Mulder et al. 2008, 2011), and that such research can be justified using arguments arising from EID and the SRK taxonomy. What remains to be seen is whether or not a system developed specifically with these considerations in mind does indeed support behaviour at the skill-based level of cognitive control, and whether auditory or visual stimuli providing equivalent information do indeed support reasoning at the rule-based level of cognitive control. Moreover, there is a question regarding whether or not supported behaviours persist after removal of the system (e.g. if the driver were to borrow another's car, one without such information). Would the driver simply learn to rely on the system, rather than associating other cues in the environment with the cues provided by the system, or would these associations occur, resulting in an implicit knowledge of the environmental (i.e. not from an eco-driving support system) cues for action? The following chapters begin to address such questions.

7 In-vehicle Information System Design

7.1 INTRODUCTION

Thus far this book has dealt with a variety of issues and has undergone a continued process of refinement, in terms of both the overall objectives, and of the theoretical backdrop that gives it direction. Attention has shifted away from the initial goal of applying ecological interface design (EID) to the issue of low-carbon vehicle interface design, towards the support of skill-based eco-driving behaviours in any road vehicle. The previous chapter presented the first major step towards this aim by offering a series of decision ladder analyses of the specific behaviours that characterise fuel-efficient driving. These analyses resulted in a discussion of the EID theoretical approach, with particular attention paid to the skills, rules and knowledge (SRK) taxonomy of behaviour. A discussion of haptic feedback in the vehicle was offered, resulting in the argument, or rather general hypothesis that combining the action and control surfaces in the vehicle (i.e. providing information to the very site on to which control actions are performed) may readily support behaviour at the skill-based level of control.

This chapter presents the next step; it provides a description of the in-vehicle system designed to investigate the theoretical arguments presented in the previous chapter. This has practical implications, in terms of the type of system that best supports fuel-efficient driving, as well as theoretical implications, in terms of whether or not the arguments for combining action and observation surfaces also follow for the combination of action and observation sensory modes (e.g. Rasmussen and Vicente 1989; Vicente and Rasmussen 1992). Moreover, there is a question of whether skill-based eco-driving behaviours can be encouraged in the novice eco-driver, and whether haptic information presented at the site of control (i.e. the accelerator pedal) does this to a greater extent than visual or auditory information (or combinations thereof). This chapter describes the construction, functioning and early pilot testing of that in-vehicle system.

7.2 THE SYSTEM

The system, capable of providing haptic, auditory and visual feedback, either individually or in any combination thereof, was designed for insertion into the Southampton University Driving Simulator (Figure 7.1). This fixed-base driving simulator used a Jaguar XJ saloon as the donor car. Three driving displays (each 241 cm wide, 183 cm high) provided the driver with a 135-degree field-of-view of the environment ahead. The driving scene to the rear was projected on to a screen

FIGURE 7.1   Southampton University Driving Simulator.

directly behind the car, viewable via the vehicle's rear-view mirror. Wing mirrors were simulated using LCD screens. Simulations were run using the STISim Drive™ M500 W Wide Field-of-View with Active Steering software system (software build 2.08.08), allowing for a 30 Hz data capture rate. The software was also able to simulate an automatic transmission; this was used for all driving sessions henceforth discussed in this book.

Following the results from, and discussions of the decision ladder analyses presented in Chapter 6, the in-vehicle information system was designed to provide an alert to the driver to encourage them to remove their foot from the accelerator pedal. The first prototype of the system comprised a vibrating pad, a box in which a light was housed and from which a sound emanated, and a control box from which it was possible to control the frequency and intensity of the vibrating pad, and the frequency of the auditory and visual stimuli (Figures 7.2 and 7.3). Note that the tonal frequency of the auditory tone was not manipulated itself, rather an unchanging tone was presented in bursts of variable length, up to continued presentation (i.e. constant sound). This was also the case for the light (brightness and colour were always the same, only the length of the bursts could be varied, up to a steady light being displayed).

Note that in Figure 7.3 though the knobs on the left (for auditory and visual stimuli) appear under the heading 'intensity' it is not the intensity of the stimuli being manipulated, rather the length of the bursts in which they are presented. Under the 'shaker' heading, the 'intensity' label is equally misleading. The knob under the 'frequency' heading does indeed control just that, the frequency of the vibrations; however, the 'intensity' knob once again controls the length of each burst of stimulus presentation, up to continuous presentation. As this box was intended only to be used by the experimenter, and not to be seen by any prospective participant, this was not considered an issue of significant concern. The 'override' button on the right of

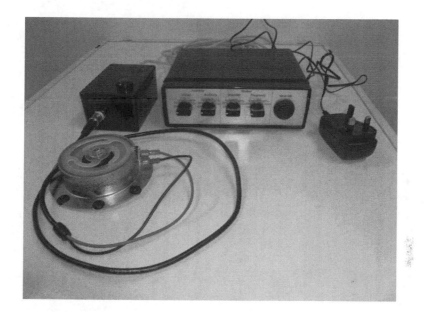

FIGURE 7.2   Initial information system prototype.

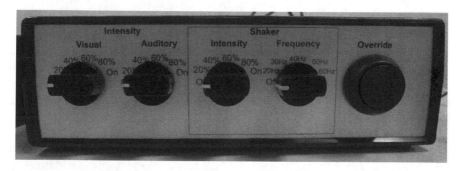

FIGURE 7.3   Close-up of control box front pane.

the control box provided the experimenter with the means to turn the stimuli on and off. This button did not require the experimenter hold it in, rather it clicked on upon depressing, and clicked off when pressed again.

To assess the physical suitability of the system, that is to say the 'feel' of the pad under foot and the possible placement of the light and sound box on the dashboard of vehicle, an initial attempt was made to integrate it into the University of Southampton Driving Simulator (Figure 7.4). It was quickly decided that the vibrating pad, which in Figure 7.4 is held in place by a simple clamp (a temporary set-up to assess the feel of the vibrator under foot), had too great a depth, that is, it excessively raised the height of the accelerator pedal. Furthermore, the nut on the top and in the centre of the pad noticeably changed the feel of the pedal. This form of vibrating pad was therefore discounted as a possibility as it would too drastically change the driver's experience.

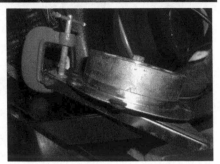

FIGURE 7.4   Information system in situ.

A different approach to the vibrating pad was therefore taken. The previous device (shown in Figure 7.4) had the capacity to vibrate at differing frequencies and amplitudes, thereby providing an additional avenue for further research. This ability, however (i.e. to independently manipulate the frequency and amplitude with which the device vibrated), was sacrificed in order to retain the required intensity of vibrations (able to be felt through the shoe of the driver) and to minimise the size of the device (to not significantly change the feel of the pedal when it is not vibrating). An array of small vibrating motors, usually used in mobile phones, was chosen as a potential solution. Six 3-volt, button-type (10 mm diameter × 3.4 mm depth) motors were attached to the rear of a metal plate, cut to the same shape as the accelerator pedal, and placed over the pedal (Figure 7.5). This raised the height of the pedal by approximately 10 mm in comparison to the brake and clutch pedals. This was considered acceptable for the purposes of this research.

The most significant change made to the control box was the functioning of the override button and the light and tone characteristics. Previously the experimenter had to switch on the stimuli, using the override button, which would be presented in bursts (of lengths dictated by the positioning of the various dials). It was decided that this type of stimulus presentation (on/off bursts) was not required. Should the driver exceed the suggested accelerator pedal depression, the information prompting them to release the pedal should be displayed continuously for the duration that they continue to depress it.

This arrangement more accurately follows the previously discussed decision ladder models; when approaching an event that requires the vehicle to slow down, the cues in the environment used by the expert would be largely constant (inasmuch as, e.g. the presence of a traffic light or a bend in the road are constant). Hence any system attempting to support these kinds of behaviours (a system which, in essence, attempts to provide an in-vehicle stimulus that is a proxy for the amalgamation of

FIGURE 7.5   Accelerator pedal with vibrating pad attached.

FIGURE 7.6   Updated control box.

environmental stimuli used by expert eco-drivers to guide behaviour) should present information continuously. Moreover, such an approach has been previously demonstrated in haptic feedback presented through the accelerator pedal for car-following support (e.g. Abbink 2006; Abbink et al. 2008; Mulder et al. 2008, 2011).

Moreover, the button itself made an audible clicking sound when operated. As this study is interested in separating out the effects of stimuli of different modes this characteristic was deemed inappropriate. In this situation there would be the possibility that the participant would respond to the sound of the button rather than the stimuli it was operating (e.g. visual or haptic information).

The control box was therefore changed so that the override button had to be depressed and held in, with the stimuli being presented continuously for the duration that the override button was depressed. When the button was released, stimulus presentation would cease. In addition to this change, the characteristics of the stimuli were altered. Figure 7.6 displays the altered control box. The dial on the left, under the heading 'visual intensity', controlled the brightness of the light. This ranged from 0% (i.e. off) to 100%, in 20 percentage point increments. The dial under the heading 'auditory tone (Hz)' was used to alter the tonal frequency of the auditory stimulus, ranging from 300 to 700 Hz in 100 Hz increments. The dial under 'shaker intensity' again controlled just that; the intensity of the vibrations provided by the array of motors, ranging from 0% (i.e. off) to 100% in 20 percentage point increments. Each motor (in the array of six) vibrated at 200 Hz and 0.8 g amplitude at 100% voltage input (see Precision Microdrives (2016) for a discussion on vibration amplitude measurements). As with all eccentric rotating mass vibrating motors, amplitude and frequency are inextricably linked, hence at lower settings (80%, 60%, 40% and 20%) both amplitude and frequency were lower. Reduction of voltage reduces frequency proportionally, whereas amplitude decreases as a square (see Precision Microdrives (2015) for more information). Finally, the override button itself was changed for one that did not make an audible sound when used.

7.3   INITIAL PILOT STUDIES

7.3.1   STIMULUS LEVELS

The control box depicted in Figure 7.6, with manual override button, was not intended as a permanent solution to the functioning of the information system (the intention was to have the simulator software itself drive stimulus presentations, as will be discussed); rather, it provided a relatively simple means for assessing the suitability of the different frequencies and intensities of the various stimuli before integrating the system into the STISim software.

An informal survey of four people, all researchers at the University of Southampton's Transportation Research Group, was conducted with the aim of assessing the effects of presenting the different stimuli at different levels of intensity (for visual and vibrational) and frequency (for auditory) on acceptance and noticeability. The aim of this exercise was to select the initial stimulus presentation settings for the pilot study. For the visual stimulus it was quickly decided to proceed with the 100% setting. This was considered most noticeable by the participants and did not present any level of annoyance or discomfort. The auditory tone was noticeable at all levels, though on average it was considered most acceptable at 400 Hz (300 Hz chosen by one participant, 500 Hz by one, 400 Hz by the remaining two). The vibrational stimulus was judged as excessively strong at and above 80% (resulting in discomfort), and not noticeable enough below 40%. Here two participants favoured 40%, two favoured 60%; a setting of 60% was chosen as the level with which to proceed with the pilot study as this was still judged to be comfortable.

As was described in the opening chapter of this book, the aim of this research project was not to offer an investigation of the effects of stimuli of differing psychophysical properties on human sensation and perception. Rather, it is an investigation of the effects of *noticeable* stimuli, presented in different modes and locations (i.e. either through the accelerator pedal or not), on performance of particular eco-driving behaviours related to use of the accelerator pedal. The choices made regarding stimulus frequencies and intensities were based purely on what was easily noticed, but not excessively annoying or uncomfortable. It is not possible to discuss the equivalence of stimuli, as they are presented in different sensory modes; they are inherently different in this respect. Noticeability is, however, important in such an investigation. Whether a stimulus is noticeable or not depends to a great extent on previous habituation to that stimulus, on the salience of that stimulus (which itself also partly depends on previous experience), and on the position from which it is presented. As can be seen in Figure 7.4, the box from which the light and sound emanated was situated on the dashboard, to the right of the steering wheel. All participants in the pilot testing phase reported being able to see the light with ease, being able to hear the sound, and being able to feel the vibration (through the accelerator pedal). The characteristics of the stimuli, as decided upon in the pilot testing phase, were therefore considered suitable for the experiments described in Chapters 8 and 9. Note, however, that the inability to guarantee equivalence of noticeability across the stimuli is still accepted as a limitation of the system, and of the research presented in Chapter 8.

7.3.2 ROUTE DEVELOPMENT AND TESTING

Due to the simulator system's limited ability to concurrently output data, the routes to be used in the main study had to be designed and created before the system could be fully tested in a pilot study. The STISim software can only output certain variables (through an RS232 port), for example distance travelled, current speed, time passed since the start, etc. It cannot provide information regarding specific events, for example corners or traffic lights, as it does not have the capacity to recognise their presence (there is no output variable to indicate such events). Hence it was necessary to specify each event in terms of the distance down the road at which it appears, and in terms of the speed down to which the driver is required to decelerate (e.g. down to 30 mph when moving from a 60 mph road section into a village, or down to a suggested 18 mph for a sharp bend in the road).

Eleven different routes, or scenarios (to use the language of the STISim software), were therefore designed. This number was chosen as it reflects the number of trials to be used in the first experimental study (reported in the following chapter).

It would have been beneficial (in terms of time and resource usage) to be able to simply modify the default routes provided by the STISim software; alas, this was not possible. Despite having scenarios set up for driving on the left-hand side of the road (necessary for a UK-based experiment), the visual appearance of the routes available in the software was biased towards a North American environment. Therefore, in order to provide road environments with a 'UK feel' (i.e. relatively narrow, single lane, winding roads, with villages of small houses, pubs, a post office and some small shops) the scenarios had to be developed entirely anew. Moreover, the simulator software did not provide any scenarios with sufficient length. For the purposes of the experiment, routes were required that matched the most common single trip distance bracket; in 2013 in the UK, 46% of car journeys made by drivers were between 2 and 5 miles in length (Department for Transport 2014). All 11 scenarios were, therefore, developed completely and solely for the purposes of the experiment reported in the following chapter.

Each scenario contained both urban and rural driving sections and simulated a drive through two villages and the surrounding rural areas in the UK (i.e. driving on

FIGURE 7.7   UK 30 mph and national speed limit (60 mph) signs.

the left, speed limits in miles per hour (mph)). All roads comprised a single lane in each direction. Each scenario was 7315 metres (24,000 feet) long in total, 2609 metres (8560 feet) of which was through villages, at a speed limit of 30 mph, with the remainder (4706 metres or 15,440 feet) at the national speed limit for single carriage roads (60 mph), representing driving on interurban, rural roads. Standard UK speed limit signs were used to indicate the 30 mph and 60 mph sections (Figure 7.7).

Each scenario had five traffic light intersections and one stop sign intersection, each of which was in a 30 mph zone (i.e. a village). Each scenario also had a simulated road accident that blocked the driver's lane. Oncoming traffic dictated that the driver slow down (to a complete stop if approaching rapidly) in order to go around the road blockage and continue along the road (i.e. they had to wait for oncoming traffic to pass before they could advance). This occurred in a 60 mph section. The scenario also had a number of road curvatures that had to be negotiated; one road curvature occurred in a 30 mph section, with a recommended cornering speed of 12 mph, five road curvatures occurred in 60 mph road sections, four of which could be safely negotiated at approximately 20 mph, one of which could be negotiated at 35 mph. Cornering at speeds faster than this would result in a simulated tyre-screeching noise (presented by the simulator software). As such, all of the road curvatures required the vehicle to slow down to lower than the posted speed limit.

Oncoming traffic was simulated throughout the scenarios at a level that would enhance realism, but not obstruct the driver (with the exception of the aforementioned road accident event). To further maximise realism, buildings, trees and pedestrians were added where appropriate. Each scenario differed in terms of the order with which events were encountered and in the types of buildings presented, and each took approximately 9 minutes to complete.

Though effort was made to have the scenarios appear different (in order to minimise learning effects and to keep the drivers more engaged in the task), it was also important that the scenarios did not differ significantly in terms of the time taken, the fuel used and the overall requirements placed on the driver.

To assess any potential differences, the 11 scenarios were driven four times, each time in a random order. All 11 scenarios were driven in each driving session, lasting approximately 2 hours and no two driving sessions occurred on the same day. For each session the order in which the scenarios were driven was randomised. As this stage of testing was concerned only with differences in routes, not with differences in drivers, it was deemed acceptable to have only one individual drive the scenarios, which was the first author of this book. This had the added benefit of not needing to train the driver in how to use the simulator (the first author had already built up considerable experience in the simulator by this point), thereby greatly reducing the potential for learning effects.

In Table 7.1 a summary of results is provided. Variables measured included the amount of fuel used (a metric provided by the simulator software), the standard deviation of acceleration rate and of the accelerator pedal position (from 0 to 1; both indicative of the smoothness of the driving profile), the standard deviation of the brake pedal force (measured in pounds) exerted across the run (also indicative of smoothness) and the total time taken to complete the run. For the fuel use, metric, the calculation performed by the simulator software (displayed in the note to Table 7.1, below) includes 'brake specific fuel consumption' as a variable. This is given in pounds mass of fuel per brake horsepower-hour (lb/bhp-hr; see, e.g. Shayler et al. 1999). The remainder of the research presented in this book uses this fuel use metric as a measure of eco-driving performance; however, it is important to state here that the validity of the solution is subject to the accuracy of the model implemented in the simulator. For each variable the score was averaged across the four runs, then converted into a z-score, that is, the number of standard deviations away from the mean (for that variable) that the figure lies. At the 5% significance level, one score is significantly different to one that would be expected assuming the routes are the same (standard deviation of brake input force, route 11). At 2.363 standard deviations from the mean (across the four runs) for that variable, this equates to a p-value of 0.018. Given that 55 observations were made, however, one would expect, based on a 5% alpha level, that one of these observations would indeed result in statistical significance. The overall results presented in Table 7.1 have therefore been accepted as indicating general equality across the 11 different routes. The different routes did not give rise to significant differences, hence we can be reasonably confident that any potential differences revealed between participants in the experimental work

TABLE 7.1
z-Score Analysis of the 11 Driving Scenarios

Route No.	1	2	3	4	5	6	7	8	9	10	11
Fuel used[a]	−0.734	0.598	−1.234	−0.901	−1.484	0.182	0.015	0.348	1.514	−0.068	1.764
StDev Accel rate (ft/s²)	−0.125	0.545	−1.145	−0.977	−1.419	0.880	0.210	−0.475	1.550	−0.642	1.596
StDev Accel position (0 to 1)	−1.745	−0.861	0.122	−0.337	−0.042	1.105	1.596	0.974	0.810	−0.369	−1.254
Brake StDev (pounds)	0.442	0.548	−0.201	0.966	−1.127	−0.310	−0.082	−0.962	−0.423	−1.213	2.363[b]
Time taken (seconds)	1.169	−0.075	−0.105	0.198	0.105	−1.766	−0.996	0.930	−1.545	1.159	0.927

[a] Fuel use was calculated by the STISim software using the following formula: Total fuel used = Total fuel used + DT × (specific fuel consumption × (9.55 × Torque × Engine speed (in revolutions per minute) / 5252)), where DT = frame rate (60 Hz) and specific fuel consumption = 0.278 lb/bhp-hr.

[b] $p < 0.05$, two-tailed.

(described in the following chapter) will not have resulted purely from differences in the 11 routes.

7.3.3 SOFTWARE INTEGRATION

As aforementioned, it was not the intention to use the box depicted in Figure 7.6 to control stimuli during the main experimental work (presented in Chapters 8 and 9), but to have the presentation timings based on simulator information and concurrent driver behaviour. This would ensure that all participants receive the same type of guidance, that is, not be subject to experimenter bias or human inaccuracies in stimulus delivery timings.

As described above, the STISim software has the capacity to output various variables in real time to an external device, in this case the control box of the information system described here. As such it was possible to provide a stimulus (or combination of stimuli – auditory, visual, haptic) to the driver based on, for example, the accelerator depression level (outputted by the STISim software as a value from 0 to 1) or the distance to a stopping or slowing event (calculated from the combination of the STISim outputs of current vehicle speed and distance to event).

Though it was not possible for the STISim software to output data that would allow the information system to automatically provide stimuli tailored to each slowing event for any scenario (e.g. there is no data marker for 'traffic light' or 'road curvature', as mentioned above), with the 11 scenarios already prepared it was possible to specify, for each scenario, how far into the run (in terms of feet travelled since the start) each slowing event occurred. This required the creation of another component to the information presentation system; one that itself contained all the required information for each scenario.

Unfortunately, the camera used to photograph this device failed, incurring the loss of all images stored. By the time of the failure's discovery the device had already been modified for use in the second experimental study (Chapter 9). An image of its original design cannot, therefore, be shown here; however, Figure 7.8 shows the modified version. The device has two rotary knobs, reflecting the design of the second experimental study (described in Chapter 9). In its first iteration, however, the device had only one rotary knob; this was to select the scenario. The device had 11 settings, one for each of the scenarios used in the first experiment. For each scenario, the distance down the road at which an event occurs was specified. Hence, rather than specifying the need to slow for a particular event (e.g. traffic light or stop sign; a piece of information that is not within the STISim software's capabilities to send) the distance down the road at which the event occurred was used (a measure that *is* outputted by the STISim software). Each scenario was therefore described in terms of the presence of each of the events, at what distance down the road they occurred (in feet), and to what speed the vehicle was required to decelerate. This information, for all 11 routes, was contained within the box depicted in Figure 7.8. Table 7.2 presents this information for scenario one. The code used to control the visual, auditory and vibrational stimuli, from the Arduino board within the control box in Figure 7.6, worked in parallel with the code for the scenario selection box (Figure 7.8) to control all stimulus timings, based on the settings on the control box (Figure 7.6), on the

FIGURE 7.8 Scenario selection control box.

TABLE 7.2
Summary of Events in Scenario One

Event	Distance From Start (feet)	Speed Limit Preceding Event	Speed Necessitated by Event
National speed limit sign	400	60 mph	60 mph (no action required)
Village – 30 mph sign	2640	60 mph	30 mph
Traffic light intersection	3590	30 mph	0 mph
Traffic light intersection	4160	30 mph	0 mph
Traffic light intersection	5280	30 mph	0 mph
Stop sign intersection	6400	30 mph	0 mph
National speed limit sign	7600	30 mph	60 mph (acceleration required)
Road curvature (moderate)	9900	60 mph	20 mph
Road curvature (moderate)	12,400	60 mph	20 mph
Roadblock (traffic accident)	14,300	60 mph	0–20 mph
Road curvature (moderate)	15,400	60 mph	20 mph
Road curvature (mild)	16,400	60 mph	35 mph
Road curvature (moderate)	17,700	60 mph	20 mph
Village – 30 mph sign	18,900	60 mph	30 mph
Traffic light intersection	19,400	30 mph	0 mph
Road curvature (severe)	20,200	30 mph	12 mph
Traffic light intersection	21,400	30 mph	0 mph
National speed limit sign	22,500	30 mph	60 mph (acceleration required)

predefined criteria (e.g. number of seconds ahead of the event, depression levels of the accelerator) and on the particular scenario being used.

From Table 7.2 it can be seen that the first event experienced by the participant was the presence of the national speed limit sign (which in this case denotes a 60 mph speed limit, in accordance with UK road laws). This required no action from the participant, rather it was to inform them of the initial speed limit for the scenario. All scenarios started in this way. Furthermore, the roadblock at (in this case) 14,300 feet into the scenario required the driver to decelerate down to different speeds, depending on the behaviour of the participant. For example, if the participant approached rapidly, they would have been required to come to a complete stop, as the oncoming traffic would not have fully passed the event by the time it was reached (i.e. the participant would have approached the event faster than the oncoming traffic, whose speed was fixed at 80 feet per second). Should the participant react early, slowing down in advance of the event, the oncoming traffic may have passed the event before the participant's vehicle came to a complete stop. Given the oncoming traffic speed, the distance down the road at which the event becomes visible, and the expected speed of the participant's vehicle preceding sight of the event, it was expected that highest speed at which the participant's vehicle was likely to be travelling would be no more than 20 mph.

From this information, and from the information concurrently outputted by the STISim software (current distance down road and current speed), it was possible to calculate a time-to-event variable to be used to dictate when stimuli encouraging deceleration were to be presented. For example, X number of seconds before event Y, present stimulus Z (i.e. auditory, visual, haptic or a combination).

To discourage excessive acceleration the process was less complex; the STISim software outputs a variable, from 0 to 1, representing the level of accelerator pedal depression. The information system simply provided a stimulus when a given threshold is exceeded (e.g. if the participant depresses the pedal more than 50%). Though a simplification, basing information on this variable is a valid method for discouraging excessive acceleration and encouraging smoother driving (see Birrell et al. 2013).

Therefore, with these two streams of information, accelerator pedal depression rate and time to event, it was possible to accurately standardise, across participants, the conditions under which they received information from the system.

7.3.4 Setting the Thresholds and Testing the System

To decide upon the exact distance to event (for stopping and slowing events) and the throttle threshold (for acceleration behaviours) to be used in the main experimental work as triggers for stimulus presentation, three individuals (the first author and the two technicians involved in developing the information system) worked collaboratively throughout the testing process. Although it would have been interesting to investigate the effect of different threshold levels on both the effect on fuel economy and on the effect on user acceptance (something that in fact occurred in the second experimental study; see Chapter 9), for the first study, fixed values were chosen. For acceleration, depression rates at or above 70% of pedal travel triggered stimulus presentation, and the time before a slowing event at which stimuli were triggered was set at 8 seconds (note that the exact point in the road at which stimuli were presented depended on the speed at which the driver was travelling).

Though relatively arbitrary, these values were considered an acceptable compromise. For acceleration, though Birrell et al. (2013) argued for (and used in their study) a 50% threshold, in the University of Southampton Driving Simulator this was deemed too conservative. At only 50% travel, acceleration was intolerably slow, and it at this depression rate it was impossible to reach speeds in excess of 40 miles per hour.

The choice of 8 seconds as a time-to-event threshold at which to present stimuli was also a compromise between expected efficiency gains and expected acceptance of the system by participants. It was thought that times greater than 8 seconds would be deemed overly conservative, and times under 8 seconds would not have had as noticeable an effect on fuel economy (the longer the coasting phase, the greater the benefit to fuel economy); however, as aforementioned, the manipulation of this value was to be the focus of the second experimental effort presented in this book (see Chapter 9).

Route testing with the system also revealed some potential problems with the deceleration advice with regard to when the advice should stop being presented. Initially, information presentation was governed by three main clauses. Stimuli would be provided if the participant

- Was within 8 seconds of the event *and*
- Was depressing the accelerator pedal *and*
- Was travelling at a speed higher than that necessitated by the event

At the traffic light intersections, it was possible to slow down in advance of the event in such a way that the light had turned green again before the vehicle had come to a complete stop. Furthermore, at the stop sign intersection, where an additional vehicle was simulated that crossed the participant's path, it was again possible for the participant to slow early enough such that the simulated cross traffic passed before the participant fully reached the intersection, thereby allowing continued travel. Though for the stop sign intersection on real UK roads this would represent an illegal manoeuvre (the vehicle is legally required to come to a complete stop at such road signage), for the sake of practicality, and with the acceptance that 'real' driver behaviour is not always strictly bound by road laws (despite what one might hope), it was decided that the speed necessitated by these events should be set to 10 miles per hour. For the roadblock incident (described above) this speed was set at 25 miles per hour for the same reasons, that is, very early action meant speeds could be in excess of 20 miles per hour once the driver was able to pass the event.

Finally, to test whether the feedback system had the desired effect (i.e. to encourage a reduction in fuel use) a pilot test with six participants was conducted. The six participants were all postgraduate researchers known personally to the current authors, and all had experience in the driving simulator. Each drove one of the routes three times (scenario two); once with no feedback and having been asked to drive normally, once with no feedback but having been asked to drive economically, and once with the feedback system in operation (with auditory, visual and haptic feedback activated in combination). For the economical trial without feedback each participant was told that economical driving is characterised by smooth driving profiles, brought about by early action to braking events (including the use of coasting), and the avoidance of heavy acceleration. For the feedback trial the participant was given a brief explanation of the system (i.e. that it encourages coasting to braking events and discourages heavy acceleration).

Each participant experienced the three trials in a different order, that is to say all the possible orders of the three trials were used. Table 7.3 presents average fuel

TABLE 7.3
Average (and Standard Deviation) Time Taken (in Seconds) and Fuel Use Across Participants

	Time (Seconds)	Fuel Use[a]
Normal	552.77 (81.57)	0.208 (0.043)
Eco	606.44 (101.45)	0.174 (0.042)
FB	647.26 (69.62)	0.156 (0.018)

[a] Fuel use was calculated by the STISim software using the following formula: Total fuel used = Total fuel used + DT × (Specific fuel consumption × (9.55 × Torque × Engine speed (in revolutions per minute) / 5252)), where DT = Frame rate (60 Hz) and specific fuel consumption = 0.278 lb/bhp-hr.

use and average time taken to complete each run. As can be seen, the averages reflect what might be expected (and hoped for given that this stage was simply meant as a means for testing the efficacy of the system). Participants used less fuel (on average) when asked to drive economically, and less fuel still when provided with information encouraging eco-driving behaviours. In Table 7.3, 'normal' refers to the trial with no feedback in which participants were asked to drive normally, 'eco' refers to the trial where participants received no feedback but asked to drive economically, and 'FB' indicates the trial in which feedback was provided. The trend in time taken is in contrast to findings from early research by Evans (1979), and Waters and Laker (1980). These studies found fuel improvements without trip time increases; however, in a controlled simulator environment involving, for example, traffic lights whose functioning is based on the car's behaviour (i.e. they will always require the vehicle to stop), the added use of coasting may well be expected to increase total trip time. This trend (i.e. more time for the FB trial, less for the Eco trial, and less still for the Normal trial) was seen across all participants. The trend in fuel use was not. Three of the six participants used less fuel in the Normal condition than in the Eco condition, and two participants used less fuel in the Eco condition than in the FB condition. Overall, however, the results from this pilot study are positive, insofar as they show a general trend of improved performance (i.e. lower fuel consumption) when asked to drive economically, and lower still when provided with eco-driving information.

Moreover, these results permit the calculation of power statistics, useful for the determination of sample size for the main experimental work. Though the pilot study sample size is too small for meaningful inferential statistics, the output can still be used to guide sample size. Pilot results were therefore subjected to a one-way, repeated-measures ANOVA (performed in SPSS version 22), from which a partial eta squared value of 0.466 was calculated. The G*Power 3.1 tool for power analysis was then used to calculate the required sample size (see Faul et al. 2007). For a one-way, omnibus ANOVA, with desired power of 0.95, an alpha value of 0.05, partial eta squared of 0.46 and with three groups (i.e. normal driving, self-guided eco-driving and driving with eco-driving feedback), the total recommended sample size is 27.

Though this sample size calculation method is based on real data (i.e. from the pilot work), it is, of course, an oversimplification. The main experimental work was not to have three conditions; rather it was to have eight experimental conditions, one base-line condition, and one 'after' condition (to test for learning effects). Furthermore, fuel use was not to be the only variable under scrutiny. Various other measures (e.g. accelerator pedal behaviour, brake pedal usage, and lane-keeping measures) were also to be investigated, alongside results from a number of questionnaires.

The G*Power software was therefore used to calculate a suggested sample size for a repeated-measures, within factors, multivariate analysis of variance (MANOVA), with 10 groups and 10 dependent variables, a power of 0.95, an alpha of 0.05, and a Cohen's f of 0.25 (Cohen's f is used by the G*Power software, and 0.25 relates to a medium effect size; Cohen (1988)). This effect size was chosen as despite the large effect shown in the pilot work (a partial eta squared of 0.46), a slightly more conservative estimate was thought appropriate. Finally, correspondence between measures was set to 0.6, a fairly conservative figure given the likely relationships between many of the measures (e.g. between average acceleration and average throttle position, or between distance spent coasting and total brake use). This resulted in a suggested sample size of 30.

7.4 CONCLUSIONS

Where the previous chapter discussed at length the theoretical foundations for the design of an in-vehicle information system for the support of eco-driving, this chapter has presented the actual design process and initial pilot testing of that system. The scenarios to be used in the first experimental testing of the system were also designed and piloted, as well as the threshold values to be used for the discouragement of excessive acceleration and the encouragement of coasting when approaching slowing or stopping events. As the reader may therefore expect, the following chapter presents the first experimental analysis of this system.

8 Ecological Driving with Multisensory Information

8.1 INTRODUCTION

In Chapter 6, the theoretical justification for the design of an in-vehicle information presentation system was offered. This drew on principles from the skills, rules and knowledge (SRK) taxonomy of human control behaviour (Rasmussen 1983), arguing that it should be possible, with haptic in-vehicle information presented at the same site on to which control actions are performed, to encourage eco-driving behaviours, at lower levels of cognitive control, in the novice eco-driver (see Chapter 6). The system itself, and the process by which it was designed and built, was described in the previous chapter. This chapter describes the first application of that system; it presents another step on the journey from theory to analysis, design and testing.

It is at this stage of the book that the two initial motivational forces (i.e. to save energy and to explore the theory underlying the SRK taxonomy and ecological interface design (EID)) truly come together. The first goal is to help drivers save fuel; the system described in the previous chapter aims to do just this, and the experiment described in this chapter tests that system. The second goal is to explore whether or not the justifications for the use of haptic information in the vehicle, made in Chapter 6, are valid. This does not present a test of the EID method; as has been discussed, the full EID process was not considered. Rather this presents an exploration of the arguments arising from the SRK taxonomy, from the direct manipulation interfaces (DMI) approach, and from a number of principles *underlying* EID (in particular, the concepts of displaying system boundaries, of supporting behaviour at lower levels of cognitive control and of allowing the user to act directly on the interface).

8.2 BACKGROUND: A RECAP

A good deal of attention has already been paid to the extant literature in justifying the design of the in-vehicle information system investigated in this chapter. The reader will be glad to hear that they will not be subjected to a full recap of all the relevant theoretical underpinnings or preceding research on the matter; however, a brief revisit *is* in order, particularly with regard to haptic feedback in the vehicle.

As previously described, in the EID literature there is a relative dearth of research that also considers information in sensory modes other than vision. Furthermore, the EID review presented in Chapter 5 highlighted the common omission of the

SRK taxonomy across the past two decades of the literature. As such, this research approaches the issue of encouraging eco-driving from a point of view that draws from both these areas; to encourage behaviour at the lowest possible forms of cognitive control (i.e. skill- and rule-based behaviour), with information that is ecologically valid in terms of the meaning it provides and in terms of the sensory mode through which it is presented.

As has been discussed, if one follows the philosophies behind DMI (Hutchins et al. 1986) and EID (Rasmussen and Vicente 1989; Vicente and Rasmussen 1992), in particular the SRK taxonomy of human behaviour (Rasmussen 1983), one could argue for the combination of action and observation surfaces (i.e. to combine the area onto which action is performed with the area from which information is obtained). Eco-driving is largely characterised by differential use of the accelerator pedal or gear stick (e.g. Barkenbus 2010; Hooker 1988). Although it is difficult to combine action and observation surfaces for gear-change advice (the main challenge being that the hand is not always resting on the gearstick, hence cannot reliably receive information from that location), it *is* possible to provide acceleration (and deceleration) advice haptically through the accelerator pedal (as has been discussed in previous chapters).

Such information systems have indeed been described in the literature, for both safety and for eco-driving support. For example, De Rosario et al. (2010) describe a vibrotactile frontal collision warning system, finding that reaction times were faster when the information was presented haptically through the accelerator pedal than when presented visually. Similarly, Mulder and colleagues (Abbink et al. 2008; Mulder 2007; Mulder et al. 2008, 2010, 2011) investigated a system that provided a counter force to the accelerator pedal when certain headway distance parameters were violated. The researchers consistently found that driver performance was improved, with a reduction in control effort compared to unsupported car-following situations (e.g. Mulder et al. 2010). This research did not focus directly on fuel use; however, not only do safe and fuel-efficient driving styles have significant overlaps (e.g. Birrell et al. 2014a), but headway maintenance behaviours themselves have been argued to have an indirect impact on fuel consumption (see Chapter 6).

Haptic feedback has also been applied to speed management. In on-road studies, Adell and colleagues investigated the use of an active accelerator pedal that provided a counter force when the speed limit was exceeded, finding that such feedback did indeed reduce instances of speeding (Adell and Várhelyi 2008; Adell et al. 2008). Furthermore, in a long-term study of the system, not only was speed compliance improved, but travel times were unaffected, and emission volumes significantly decreased (Várhelyi et al. 2004).

Research specifically looking at the effect of haptic feedback on eco-driving is less common; however, there are some examples in the literature. Hajek et al. (2011) described an active accelerator pedal that alerted drivers to upcoming driving events, encouraging removal of the foot from the pedal in order to maximise the coasting phase of the vehicle. This resulted in a 7.5% decrease in fuel consumption (Hajek et al. 2011). Birrell et al. (2013) investigated vibrotactile feedback aimed at discouraging excessive acceleration, drawing on research arguing that depression of the acceleration pedal beyond 50% of travel can be regarded as inefficient. The researchers found significant reductions in excessive throttle use in participants provided with a vibrotactile stimulus.

Jamson et al. (2013) investigated a similar concept in their investigation of an accelerator pedal that provided either force or stiffness feedback to encourage an 'idealised' (in terms of fuel-efficiency) accelerator pedal position for acceleration and cruising situations. When provided with haptic information, participants more closely followed 'ideal' accelerator pedal profiles.

Azzi et al. (2011) also investigated the possibility of supporting efficient acceleration profiles with haptic information, obtaining similar results. Further to assessing haptic information, an additional comparison with visual information was made. Though both modes supported eco-driving behaviours equally, control activity was significantly lower in conditions that included haptic information (Azzi et al. 2011). Similarly, Staubach and colleagues describe a series of studies investigating the efficacy of visual, haptic and visual-haptic interfaces for acceleration and gear-change advice, and to encourage coasting (Staubach et al. 2012, 2013, 2014a, 2014b). They found that acceleration pedal profiles and gear-change timings were closest to optimal when participants were provided with a pairing of visual and haptic information (Staubach et al. 2012).

Finally, Jamson and colleagues extended their previous work (cited above) by comparing a variety of haptic, visual, and visual and auditory interfaces (Hibberd et al. 2015; Jamson et al. 2015). All systems were found to be effective in encouraging efficient driving behaviours; however, the haptic systems were judged to best guide acceleration behaviours. Additionally, complementary auditory alerts reduced distraction under visual feedback conditions. Such a finding confirmed earlier research indicating that complementary auditory stimuli add benefit to visual displays (Kim and Kim 2012).

The study described in this chapter builds on the body of research outlined above. As aforementioned, visual information has been compared with haptic, and pairings involving these two modes (Hibberd et al. 2015; Jamson et al. 2015; Staubach et al. 2014b), and visual information alone has been compared with an auditory and visual pairing (Jamson et al. 2015; Kim and Kim 2012); however, a comparison of information with equal content (i.e. in support of the same behaviours) across all three sensory modes, and the various possible combinations thereof, is currently lacking. Furthermore, with the notable exception of Birrell et al.'s (2013) work, the vast majority of research on haptics in the vehicle investigates force or stiffness feedback rather than vibrotactile. The research presented in this chapter also addresses this gap in the literature. Gear-change advice was not investigated for a number of reasons. First, to reduce complexity such that differences in fuel use can be attributed only to acceleration behaviours; second, so that results can be generalised to vehicles without gears, for example, vehicles with automatic transmission, and vehicles with non-conventional drivetrains (including electric vehicles); and third, to allow for the comparison of information that combines action and observation surfaces with information *of equal content* that does *not* combine such surfaces. This would not have been possible for gear-change advice (see above); hence it was omitted from the study.

Although EID and the SRK frameworks have both been discussed at length in this book, it is important to reiterate, once again, that the eco-driving system described in the previous chapter, and assessed in this chapter, was *not* designed using the full EID process. A work domain analysis, one of the fundamental tools of EID, was not performed, and the system itself does not support behaviour at all three levels of cognitive control – a necessary condition for an ecological interface. However, arguments arising from the method's theoretical foundations were central to the development of the system. In particular, the system was developed primarily to satisfy the first of EID's three core principles; to support interaction via time-space signals, thereby encouraging behaviour at the lowest possible level of cognitive control (i.e. skill-based behaviour).

In Chapter 6 a series of decision ladders were presented, the analyses of which led to the conclusion that when decelerating it is in the timing of the foot's removal from the accelerator pedal that is important for eco-driving. For acceleration activities it is partly from the force with which the pedal is depressed that differences in fuel use arise. The system developed for this study therefore aims to inform the driver of the optimum levels and timings of accelerator pedal usage. The system provides alerts that aim to provide the novice eco-driver with information that allows them to take shortcuts through the decision ladder; where an expert will amalgamate various cues from the environment, drawing from their experience to guide their behaviour, the question here is whether the same shortcuts can be induced in novices with in-vehicle information.

8.3   EXPERIMENTAL AIMS

The main aim was, of course, to assess the effectiveness of a system designed to support eco-driving in the novice eco-driver. Hence perhaps the most important question to ask is; does it work, and would people use it? In other words, does the system help drivers to use less energy (and display more associated eco-driving behaviours) while also receiving high levels of acceptance from the participants? To this end it was specifically hypothesised that haptic and auditory information, and combinations of stimuli containing these modes, would foster greater compliance than visual information, but that visual information would be more accepted by participants (in line with findings from Staubach et al. 2014a).

Another hypothesis can be formulated with regard to workload. Not only is driving a largely visual task (Spence and Ho 2009), but the in-car environment is replete with visually dominant in-vehicle technologies. As Harvey et al. (2011b) argued, one of the main priorities for any in-vehicle information system must be to minimise conflicts with the primary driving task, thus reducing the likelihood of distraction. Considering work on the multiple resource theory (e.g. Wickens 2008) one could argue that increasing demand on the visual channel will have a more detrimental effect on driving performance, in terms of workload and distraction, than will providing information with equal meaning through the auditory or haptic channel. Considering this postulation, alongside results from the aforementioned study reported in Birrell et al. (2013), it was hypothesised that visual information would have a stronger, negative effect on overall driving performance and workload than would auditory or haptic information.

Finally, following on from the arguments forwarded by the proponents of EID (Vicente and Rasmussen 1992), the SRK taxonomy (Rasmussen 1983) and the DMI approach (Hutchins et al. 1986), and expanded upon in this book, one could argue that haptic information, presented at the site of control, is more likely to exert influence over behaviour at the skill-based level of cognitive control than are visual or auditory stimuli. This question is, therefore, whether or not haptically presented eco-driving information (in the form of a vibrotactile stimulus), presented at the site of control better supports eco-driving behaviours than information with equivalent content presented through the auditory or visual channel.

Leading on from this, if one considers that skill-based behaviour progresses largely outside conscious awareness, (see Rasmussen 1983, 259), it might be expected that previously held beliefs and habits play less of a role in influencing behaviour at this level of cognitive control than when interacting at the rule- or knowledge-based level of cognitive control, where there is a more conscious interpretation of the stimuli–response pairing. The cognitive control behind a participant's response will not include the considerations of goals and options as presented in the top section of the decision ladder (or at least be occurring in such a way as to be outside the actor's conscious awareness). Should this be the case, that is, should it be largely skill-based behaviour that is encouraged by the information presented by the current study's eco-driving information system, similar behaviour change should be displayed by all participants, regardless of their opinions on fuel use.

One way to investigate this is to first assess an individual's general attitude towards the environment and the issue of climate change. Should an individual be more environmentally aware, and be more concerned about our effect on the planet we inhabit, it is more likely that they will exhibit eco-driving behaviours once informed about the impact driving style has on the use of fuel and the resultant greenhouse gas emissions (see Chapter 2 for a discussion on activating specific goals and their effect on behaviour, and Chapter 3 for a discussion of the link between environmental awareness and eco-driving knowledge and behaviours). If the individual is less concerned about their effect on the environment in which we live, or simply does not consider climate change to be as significant as suggested by the IPCC (e.g. Intergovernmental Panel on Climate Change 2007), they may be less likely to exhibit such eco-driving behaviours. There is, therefore, the possibility that those participants with stronger positive attitudes towards the environment (i.e. more environmentally conscious) will generally drive in a more fuel-efficient manner when behaviour is driven at the knowledge-based level of cognitive control, and, relatedly, that if the information system under investigation in this study supports eco-driving at the skill-based level of control, it should do so equally among all participants, regardless of their underlying attitude towards the environment.

These final issues are, however, rather exploratory in nature, and have therefore not been formally described as specific hypotheses, but left as more open suggestions. Indeed, in the words of Jens Rasmussen, 'the boundary between skill-based and rule-based performance is not quite distinct, and much depends on the level of training and on the attention of the person' (Rasmussen 1983, 259).

8.4   METHOD

The experimental study used a repeated-measures design, with both within- and between-subject measures. One independent variable, containing 10 levels, represented the within-subject factor. Each of the 10 levels related to a driving condition in which different combinations of auditory, haptic and visual information were presented, including three conditions in which no additional information was presented (summarised in Table 8.1 in the procedure section, below). The order in which participants experienced the various trials is described below. Four between-subject factors were investigated; environmental attitudes, driving experience, driving level and baseline fuel consumption. For each factor participants were split into two groups; these are described in more detail below, and summarised in Table 8.2 (also in the procedure section). Table 8.2 also summarises the 12 independent variables used in the study; these covered various aspects of driving performance, system acceptance and self-reported workload.

8.4.1   PARTICIPANTS

Thirty participants, 17 male, 13 female, ages 23 to 59 (M = 33.83, SD = 11.95), were sought via a convenience sample. Participation was entirely voluntary and all participants provided fully informed consent; none were paid for their time. Ethical approval was sought from, and granted by, the University of Southampton's Ethics and Research Governance committee, reference number 10612.

8.4.2   APPARATUS

8.4.2.1   Driving Simulator

All driving sessions took place in the STISim-equipped Southampton University Driving Simulator, described in detail in Section 7.2. The previous chapter describes in detail the system that was used to provide all necessary stimuli; nevertheless, a brief recap is offered here.

The system provided vibrational feedback via an array of six 3-volt vibrating motors attached to the back of a metal plate. These vibrated at 12,000 RPM (200 Hz) with 0.8 g amplitude. The metal plate was secured on to the top of the accelerator pedal (see Figure 8.1) – this raised the height of the foot position by approximately 10 mm compared to the brake and clutch pedal; this was deemed acceptable in pilot studies.

The box shown in Figure 8.2 provided both the visual and auditory stimuli, and was positioned on the dashboard above and to the right of the steering wheel, the positioning of which was deemed suitable in pilot studies, that is, within the driver's field of view, but not obstructing the view of the road environment (Figure 8.3). The light at the top right of the box provided a red light, and the auditory tone emanated from the speaker at the top left of the box, with a frequency of 400 Hz. Note that the priority in this study was not to investigate the driver's psychophysical responses to various stimuli intensities; the characteristics of the stimuli (auditory, visual and haptic) were chosen as they were judged to be noticeable and deemed acceptable by participants in pilot studies (see Section 7.3.1).

FIGURE 8.1 Accelerator pedal with vibrating pad attached. See also Figure 7.5, Chapter 7, this volume.

FIGURE 8.2 Box from which the visual and auditory stimuli were presented.

8.4.2.2 Driving Scenarios

As has been described in detail in Section 7.3.2, 11 different scenarios (i.e. road environments) were developed for the study, each of which simulated a UK environment including both urban and rural driving sections. These will not be described again here (the reader is referred to Section 7.3.2), suffice it to say that each scenario contained an equal number of identical events, were of identical lengths and were judged to be sufficiently equal in the demands placed on the driver (again, see Section 7.3.2).

8.4.2.3 Information System Functioning and Data Capture

As described in Chapter 7, to discourage excessive acceleration a 70% acceleration pedal depression threshold was set. Though Birrell et al. (2013) used a 50% threshold

FIGURE 8.3 Box presenting visual and auditory stimuli, positioned above and to the right of the steering wheel.

in their research, this was identified as too conservative in pilot studies. At this rate the vehicle was unable to reach the 60 mph speed limit at any point in the trials. As a compromise between conservatism and fuel-economy a 70% threshold was chosen, having been judged to be suitable in pilot testing. The system therefore provided a stimulus when, and for as long as, the accelerator was depressed further than 70% of its travel.

When approaching a deceleration event, a lead time of 8 seconds was chosen; that is to say, at 8 seconds before a stopping event the system would provide the stimulus. Stimuli were provided only if the accelerator pedal was depressed and the vehicle was travelling faster than the target speed (i.e. the speed necessitated by the event). Stimulus presentation would stop as soon as the accelerator pedal had been released, as soon as the target speed had been reached, or as soon as the event had been passed. For traffic lights and stop signs this target speed was 0 mph, while for road curvatures this was 35 mph, 20 mph or 12 mph, depending on the specific road curvature in question. For the simulated roadblock this was 25 mph, as given early action to the event (i.e. to remove the foot from the accelerator as soon as the event was in view) it was possible to reach it at this speed after the oncoming traffic had passed, hence allowing the driver to move into the right-hand lane to go around the blockage. For speed limit changes this threshold was set to 30 mph.

It is worth noting that these lead distances were sufficiently far from the corners and road blockage that simply coasting would allow the vehicle to come down to the required speed; however, for traffic lights and stop signs the participants were still required to apply the brakes (though only minimally) in order not to travel past the event.

8.4.2.4   Questionnaires

The NASA-TLX (Hart and Staveland 1988) was adopted as a means for assessing participants' workload. This six-item questionnaire elicits subjective ratings, on a 20-point scale from low to high, on perceived mental, physical and temporal demands, frustration, effort and performance. The original 1988 paper describing the method also employs a weighting process; however, in this research this has not been included. As such, this presents an application of the raw TLX. Not only is this easier to apply, but it requires less time to complete and has been shown to be equally valid (see Hart 2006 for a review). The raw TLX can be found in Appendix A.

To assess participants' subjective ratings of the information system and its various presentation methods the Van Der Laan acceptance scale was used (Van Der Laan et al. 1997). This nine-item questionnaire elicits responses from participants on a 5-point scale, measuring both perceived usefulness of, and general satisfaction with the system under evaluation. It was developed as a means for specifically assessing the acceptance of advanced transport telematics and has been used in a variety of studies since its inception (e.g. Cocron et al. 2011; Kidd 2012; Shahab and Terken 2012; Skoglund and Karlsson 2012), including studies on eco-driving advice systems (e.g. Staubach et al. 2013). The questionnaire can be found in Appendix B.

For the assessment of an individual's general attitude towards the environment the environmental attitude inventory was used (Milfont and Duckitt 2010). The full questionnaire presents participants with 193 items, each of which invites a response on a 7-point scale, ranging from strongly agree to strongly disagree, around half of which are reverse-scored. The full version of the inventory was considered too long for the purposes of this study; however, the originators of the method also provide two shortened versions of the questionnaire, the 'short' version, consisting of 72 items, and the 'brief' version, consisting of 24 items (Milfont and Duckitt 2010). It was the short version that was used in this study. This can be found in Appendix C. Although the questionnaire can be divided into 12 subscales, it is also possible to calculate a global score, ranging from 72 to 504 (lower to higher pro-environmental attitudes). The present research made use of this global score.

8.4.3 PROCEDURE

Each participant was subjected to 11 driving trials, each of which lasted around 9 minutes. The first of these was a training run, in which each participant drove the same scenario. They were told that this was simply to 'get used to' the simulator. The particular scenario used in each of the remaining 10 experimental trials was randomised across participants. For the first experimental trial, the baseline trial, the participants were asked to drive 'as they normally would'. No indication was given that the study was interested in eco-driving behaviours and the participants were not provided with any feedback. Immediately upon trial completion the participants were provided with the Van Der Laan acceptance scale and the NASA-TLX questionnaire. For the Van Der Laan scale, participants were specifically told to rate the information currently in the vehicle (i.e. the information already present in the vehicle's dashboard). The participants were also asked to complete the environmental attitude inventory at this point in the experiment.

The second experimental trial, the 'eco' trial, was identical to the first, with the exception that participants were now fully informed of the eco-driving focus of the study. Furthermore, for this trial they were asked to drive in an economical manner.

They were not told how this was to be achieved. The Van Der Laan acceptance scale and the NASA-TLX questionnaire followed completion of the trial. Again, participants were asked to complete the Van Der Laan scale with the standard in-vehicle information in mind (e.g. speedometer, tachometer).

The next seven trials each involved some form of eco-driving feedback, either visual, auditory, haptic or any combination thereof. Before the first of these trials (i.e. before trial 4 in Table 8.1, below) the participants were briefed on the functioning of the system. They were told that they would receive information suggesting that they were either depressing the accelerator pedal excessively, or that they were depressing it at a time when a coasting phase was suggested. They were expressly informed that the same type of stimulus would be used to guide both behaviours, and that presentation of the stimulus could mean that they should lift the foot off entirely, *or* that they should release it only partially (i.e. to return below 70% depression). The participants were not told which information modality would be used before each trial.

Due to the large number of different feedback combinations, true counterbalancing of the seven feedback trials was not possible (this would have required 5040 participants). The order was therefore randomly chosen for each participant; if the randomly chosen order repeated one that another participant had already experienced, a new random order was generated. The participants were asked to complete both the Van Der Laan acceptance scale and the NASA-TLX questionnaire after each of these trials. The final trial was identical to the first experimental run; no feedback was given, and the trial was again followed by the two questionnaires used in all the previous trials. The participants were not given any further instruction

TABLE 8.1
Summary of Procedure

Day	Trial Number	Description	Feedback	Route
	1	Simulator training	None	Same for all participants
1	2	Baseline trial	None	
	3	Experimental trial – participants informed of eco-driving focus and asked to drive economically	None	
	4	Experimental trial	Any of the following;	Randomised across participants
	5	Experimental trial		
	6	Experimental trial	V, A, H,	
	7	Experimental trial	V+A,	
2	8	Experimental trial	V+H,	
	9	Experimental trial	A+H,	
	10	Experimental trial	V+A+H	
	11	Learning effect assessment	None	

Note: + denotes a combination of the feedback types indicated; A, auditory; H, haptic; V, visual.

before starting this trial (i.e. they were not told that the additional information would no longer be presented). The final trial was identical to the first experimental run; no feedback was given, and the trial was again followed by the two questionnaires used in all the previous trials. The participants were not given any further instruction for this trial. The trials are summarised in Table 8.1. Finally, a short interview took place at the end of the study to gather general opinions about the system, overall experiences and design preferences.

Due to the length of the experiment (approximately 3 hours in total), and the resultant possibility of fatigue or simulator sickness, it was split across 2 days. The participants completed the training run and the first five experimental runs on day one, and the remaining five runs on day two. Every effort was made to have day two follow immediately from day one (i.e. consecutive days); however, this was not possible for four of the participants. For three there was a gap between sessions of 2 days, and for one there was a gap of 3 days. Due to time constraints one participant completed the study across three sessions.

The study's primary independent variable was the experimental condition, that is, the type of in-vehicle information experienced, and had 10 levels (trials 2 to 11 in Table 8.1, above). However, given the further aim of investigating the potential differences between different groups of participants, a number of additional independent variables were used, calculated from the data. These are summarised in Table 8.2.

For experience and driving level, the participants were split into two groups based on their responses to the questionnaires; however, as some participants gave identical responses (either in years for experience, or in miles per year driven for driving level), in order to meaningfully perform such a split, the sizes of the two groups were not the same for these variables. For experience, those with less than 9 years' experience totalled 14 individuals, with 16 individuals reporting more than 9 years' experience. For driving level, 17 individuals reported driving more than 8000 miles per year; 13 reported driving less than this figure. For environmental attitude and baseline fuel group, a true median split was possible; hence the two groups compared for these variables each contained 15 participants.

Table 8.2 also summarises the various dependent measures used to assess driving performance, self-reported workload and subjective user acceptance.

8.5   RESULTS

Due to the non-parametric nature of the data, Van Der Laan usefulness and satisfaction scores and NASA-TLX scores were analysed using Friedman's test, with post-hoc analyses performed using Wilcoxon signed-rank tests (see, e.g. Field 2009). Due to unacceptable violations of normality (a necessary pre-condition for the use of parametric tests; see, e.g. Field 2009), the 'time over 0.7' and 'excess acceleration' variables were also analysed using these non-parametric tests. As all remaining variables met the requirements for use of parametric statistical analyses, a repeated-measures multivariate analysis of variance (MANOVA), with both within- and between-subjects factors, was applied (again, see Field 2009 for discussions on the use of appropriate statistical analyses).

TABLE 8.2
Summary of Independent and Dependent Variables

Variable Type	Variable Name	Variable Description
IV	Feedback method	Within-subjects factor; experimental condition (see Table 8.2 for more detail)
	Environmental Attitude	Between-subject factor; participants split into two groups (low, $n = 15$, and high, $n = 15$) based on their responses to the EAI
	Experience	Between-subject factor; participants split into two groups (less than 9 years, $n = 14$, and more than 9 years, $n = 16$) based on the number of years since obtaining a driving licence
	Driving level	Between-subject factor; participants split into two groups (8000 or more miles per year, $n = 17$, and less than 8000 miles per year, $n = 13$, and more than 9 years) based on annual mileage
	Base fuel use [a]	Between-subject factor; participants split into two groups (high, $n = 15$, and low, $n = 15$) based on their fuel usage in the baseline trial
DV	Time taken	Time taken to complete each trial
	Fuel use [a]	Fuel used in each trial
	Throttle mean	Mean throttle position (0 to 1) across trial
	Throttle max	Maximum throttle position across trial (0 to 1)
	Throttle SD	Throttle position standard deviation across trial (0 to 1)
	Brake SD	Standard deviation of brake pedal input force (in pounds)
	Distance coasting	Total distance spent travelling forward (i.e. >0 mph) without depressing throttle
	Brake use	A measure of overall brake use, calculated by taking the total area under the curve created by brake pedal input force by time across entire trial
	Excess acceleration	Product of the magnitude of throttle position (when over 70% of depression) and time spent over the 70% threshold
	Satisfy	Satisfaction score on the Van Der Laan acceptance scale
	Useful	Usefulness score on the Van Der Laan acceptance scale
	NASA-TLX	NASA-TLX workload score

Note: EAI, environmental attitude inventory; see text, above, for a description; NASA-TLX, NASA task load index; see text, above, for a description.

[a] Fuel use was calculated by the STISim software using the following formula: Total fuel used = Total fuel used + DT × (Specific fuel consumption × (9.55 × Torque × Engine speed (in revolutions per minute) / 5252)), where DT = frame rate (60 Hz) and specific fuel consumption = 0.278 lb/bhp-hr.

8.5.1 Objective Measures

A Friedman test revealed significant differences between conditions for the measure of excess acceleration ($X^2(9) = 102.857$, $p < 0.001$). Wilcoxon signed-rank tests (with the Bonferroni-Holm correction applied) were again used to investigate pairwise differences. Group means are presented in Figure 8.4. The figure indicates that when asked to drive economically participants display fewer harsh acceleration behaviours than when asked to drive 'normally', and that when provided with

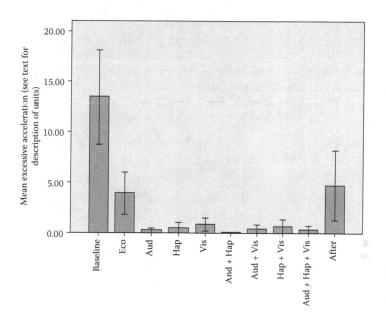

FIGURE 8.4 Mean excessive acceleration figures, with 95% confidence intervals.

eco-driving information this harsh acceleration is further reduced. The baseline trial differed significantly from all others (at the 5% alpha level), and the eco trial differed from all feedback trials except the visual and visual/haptic trials (again, at the 5% alpha level). Furthermore, these effects persisted once the additional information had been removed; though participants were more likely to show harsh acceleration behaviours in the after trial than in feedback trials, they did not fully return to their baseline performance (results in the after trial differed significantly from all others, except the eco trial).

A final point to note regards the comparison of the visual trial with the auditory/haptic and auditory/haptic/visual trials. Visual information was significantly less effective at reducing harsh acceleration than were trials involving auditory, or auditory and haptic information together (once again at the 5% alpha level).

For the remaining objective measures, in the mixed-model MANOVA, Box's test of equality of covariance matrices could not be computed as there were fewer than two non-singular cell covariance matrices. Accordingly, the Pillai's trace omnibus statistic (V) was used in preference to Wilks' lambda. All variables except distance coasting and throttle max violated Mauchly's test of sphericity. As such, results are reported in terms of the Greenhouse-Geisser correction. The multivariate analysis revealed a significant within-subjects main effect for treatment (i.e. type of feedback trial), $V = 1.298$, $F_{(72,1152)} = 3.100$, $p < 0.001$, partial $\eta^2 = 0.161$. No significant main effects were found for between-subject factors. The only significant interaction effect was between base fuel group and the effect of the different feedback conditions on the fuel use variable, $V = 0.701$, $F_{(72,1152)} = 1.538$, $p = 0.003$, partial $\eta^2 = 0.088$.

Subsequent univariate analyses of variance tests revealed significant differences between groups for all dependent measures; results for the main effects are reported in Table 8.3. Data for all variables are presented in Figures 8.5 to 8.12, and include the results of the post-analyses of variance (ANOVA) pairwise comparisons. Comparisons significant at the 5% level (after Bonferroni-Holm corrections) are displayed.

As can be seen from Figures 8.5 to 8.9, a similar pattern of results was observed for the majority of measures. Upon asking participants to drive in a fuel-efficient

TABLE 8.3
Summary of Significant ANOVA Results for Within-subjects Main Effect

Dependent Variable	F (df)[a]	p Value	Partial η^2
Time taken	4.870 (4.696,75.143)	< 0.001	0.233
Fuel use	19.628 (2.202,35.229)	< 0.001	0.551
Throttle mean	21.159 (3.925,62.797)	< 0.001	0.569
Throttle max	13.242 (5.635,90.166)	< 0.001	0.453
Throttle SD	6.859 (3.648,58.370)	< 0.001	0.300
Brake use	2.806 (3.258,52.133)	0.044	0.149
Brake SD	4.218 (3.488,55.806)	0.007	0.209
Distance coasting	20.961 (5.703,91.240)	< 0.001	0.564

[a] df, degrees of freedom, with Greenhouse-Geisser correction applied.

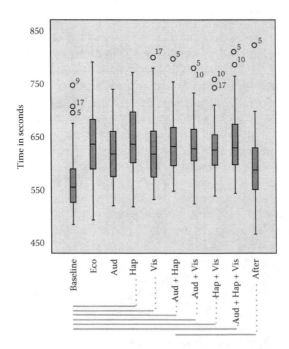

FIGURE 8.5   Time taken (seconds), by condition. Lines below graph indicate all significant pairwise comparisons.

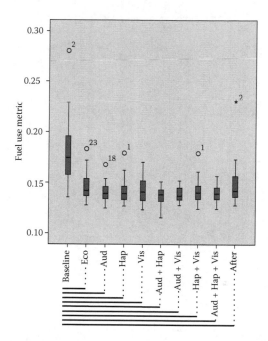

FIGURE 8.6 Fuel used (metric calculated by simulator software; see Table 8.2) by condition. Lines below graph indicate all significant pairwise comparisons.

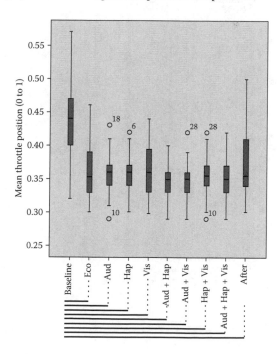

FIGURE 8.7 Mean throttle position (0 to 1), by condition. Lines below graph indicate all significant pairwise comparisons.

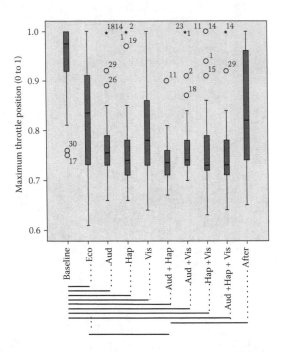

FIGURE 8.8 Maximum throttle position (0 to 1), by condition. Lines below graph indicate all significant pairwise comparisons.

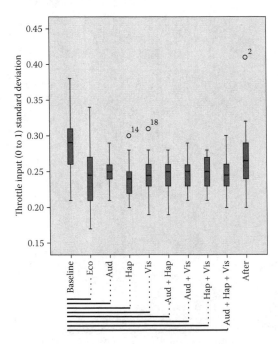

FIGURE 8.9 Throttle position (0 to 1) standard deviation, by condition. Lines indicate below graph all significant pairwise comparisons.

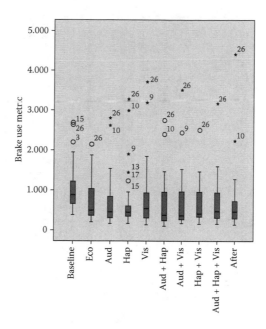

FIGURE 8.10 Brake use metric [the area under the curve created by brake pedal input (pounds) by time, over the entirety of the trial (see Table 8.2)], by condition. No significant pairwise comparisons found.

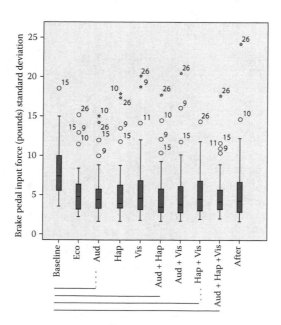

FIGURE 8.11 Brake pedal input (pounds) standard deviation, by condition. Lines below graph indicate all significant pairwise comparisons.

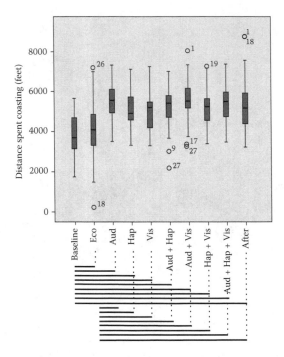

FIGURE 8.12 Distance spent coasting (feet), by condition. Lines below graph indicate all significant pairwise comparisons.

manner, they significantly changed their driving behaviour. Although they continue to drive in a fuel-efficient manner under conditions of feedback, there were no significant differences in performance under feedback conditions compared to the eco condition (with one exception; mean throttle position was significantly lower in the auditory/haptic trial than in the eco trial). Generally, this effect persisted into the after trial; however, although the pattern of results is similar across Figures 8.5 to 8.9, after trial performance was *significantly* different from baseline only for the fuel use and mean throttle position metrics.

Figures 8.10 and 8.11 (regarding brake use) show similar patterns; however, they are less marked, with fewer statistically significant results. As with Figures 8.5 to 8.9, results for the distance coasting variable (Figure 8.12) also show significant differences in performance between baseline and eco conditions; however, significant differences were also observed between the 'eco' and all of the feedback trials. This effect persisted into the 'after' trial in which no information was provided. Participants travelled significantly further using only the momentum of the vehicle (without the need for accelerator pedal depression) when asked to drive economically, and significantly further still when provided with feedback encouraging them to do so.

There were very few differences between the feedback trials for any of the variables investigated. Although not statistically significant, a trend can be seen regarding the visual-only trial, particularly in Figures 8.6–8.8 and 8.11. Results were more varied in this trial than in other feedback modes and combinations. This suggests that, compared to modalities involving auditory or haptic information, visual-only information encourages compliance to a lesser extent in some participants. This pattern can also be seen in Figures 8.13 to 8.16 below, those depicting interaction effects.

As aforementioned, a significant interaction effect was found for base fuel group. Subsequent univariate analyses revealed significant effects for the following independent measures; time taken ($F_{(4.696, 75.143)} = 4.402$, $p = 0.002$, partial $\eta^2 = 0.216$), fuel used ($F_{(2.202, 35.229)} = 5.723$, $p = 0.006$, partial $\eta^2 = 0.263$), mean throttle position ($F_{(3.925, 62.797)} = 7.162$, $p < 0.0005$, partial $\eta^2 = 0.309$) and standard deviation of throttle position ($F_{(3.648, 58.370)} = 3.287$, $p = 0.02$, partial $\eta^2 = 0.170$). No interaction effects were found between feedback condition and the participants' environmental awareness, their driving experience (in years since the awarding of their licence), or their driving level (in miles per year).

As a general finding, the differences in driving behaviour (as measured by the metrics used in this study) between the two groups when under feedback conditions are minimal. When no feedback was provided, the differences were considerably larger. At baseline results for the higher baseline fuel use group were significantly greater than those for the lower fuel use group. These differences persisted into the eco trial; however, upon the introduction of feedback, these differences disappeared. Although differences were once again greater in the after trial (the higher fuel use group were more likely to go back to a driving style that more closely resembled their baseline performance), these differences were still not statistically significant. Finally, it is worth drawing attention to the difference in results between the two groups in the visual-only trial. Although only a general pattern (i.e. not statistically

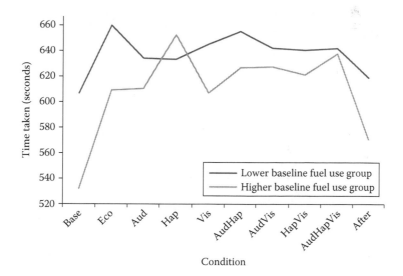

FIGURE 8.13 Interaction effect between feedback condition and base fuel group for time taken, $F_{(4.696, 75.143)} = 4.402$, $p = 0.002$, partial $\eta^2 = 0.216$. Dark grey line shows performance, under the various conditions, of those that used less fuel in the baseline condition; light grey line shows performance of those that used more fuel in the baseline condition.

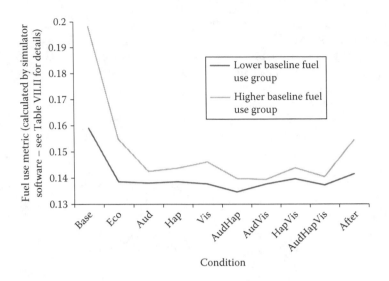

FIGURE 8.14 Interaction effect between feedback condition and base fuel group for the fuel use variable, $F_{(2.202, 35.229)} = 5.723$, $p = 0.006$, partial $\eta^2 = 0.263$. Dark grey line shows performance, under the various conditions, of those that used less fuel in the baseline condition; light grey line shows performance of those that used more fuel in the baseline condition.

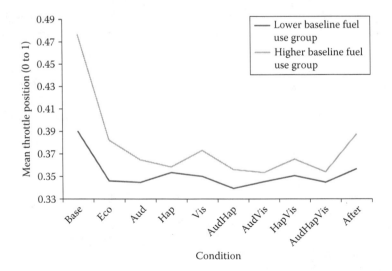

FIGURE 8.15 Interaction effect between feedback condition and base fuel group for the throttle mean variable, $F_{(3.925, 62.797)} = 7.162$, $p < 0.001$, partial $\eta^2 = 0.309$. Dark grey line shows performance, under the various conditions, of those that used less fuel in the baseline condition; light grey line shows performance of those that used more fuel in the baseline condition.

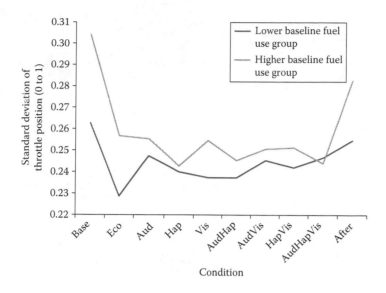

FIGURE 8.16 Interaction effect between feedback condition and base fuel group for the throttle SD variable, $F_{(3.648,58.370)} = 3.287$, $p = 0.02$, partial $\eta^2 = 0.170$. Dark grey line shows performance, under the various conditions, of those that used less fuel in the baseline condition; light grey line shows performance of those that used more fuel in the baseline condition.

significant), it can be seen that in this trial the higher fuel use group displayed a slight return to baseline performance whereas the lower fuel use group did not. This reflects the patterns seen in Figures 8.6–8.8 and 8.11 above.

8.5.2 Subjective Measures

Friedman tests revealed significant differences between conditions for Van Der Laan usefulness scores ($X^2(9) = 65.973$, $p < 0.001$), for satisfaction scores ($X^2(9) = 89.505$, $p < 0.001$) and for NASA-TLX workload scores significant ($X^2(9) = 27.684$, $p = 0.001$). Group means (with 95% confidence intervals) are presented in Figures 8.17 and 8.18.

Pairwise comparisons were performed using Wilcoxon signed-rank tests for each of the three variables presented below. All discussions of statistical significance herein are in reference to two-tailed significance tests at the 5% alpha level, after application of the Bonferroni-Holm correction for multiple comparisons.

The only significant pairwise comparisons for the usefulness variable (Figure 8.17a) involved either the eco condition or the after condition. The haptic, auditory/haptic, and haptic/visual information presentation methods were all rated as significantly more useful than information given in the eco condition (i.e. no additional information, only that from the dashboard and driving environment). The baseline trial did not yield ratings that were significantly different from other trial ratings, despite displaying information that was the same as in the eco trial. In the after trial the system was rated significantly less useful than in any of the feedback trials, with the exception of visual feedback alone.

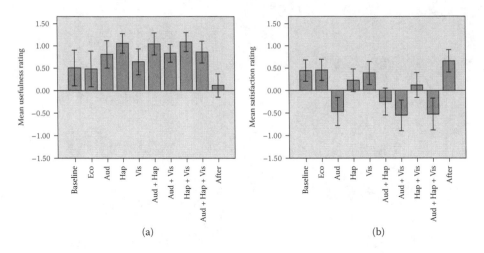

FIGURE 8.17   Mean Van der Laan satisfaction (a) and usefulness (b) ratings, with 95% confidence intervals.

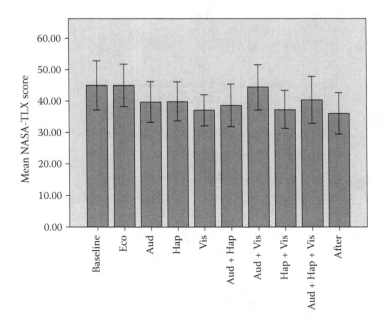

FIGURE 8.18   Mean NASA-TLX scores, with 95% confidence intervals.

With regard to satisfaction ratings it can be seen from Figure 8.17b that the information given in trials that involved an auditory stimulus (alone or in combination) was rated as consistently less satisfying than information given in trials that did not include an auditory tone (or indeed a lack of additional information). All comparisons of the auditory stimulus (alone or in combination) with combinations *not* involving the auditory stimulus were significant, with the exception of the comparison between the haptic and auditory/haptic trials (though the trend is still evident; Figure 8.17). No significant differences were found for any pairings *not* involving auditory the stimulus.

No significant differences between any conditions for NASA-TLX scores were found. The baseline and eco trials appear to have attracted slightly higher workload ratings than other trials, and the after trial slightly lower ratings (Figure 8.18); however, these trends are not clear, hence do not invite definitive conclusions.

8.6 DISCUSSION

This chapter has described a driving simulator study in which 30 participants drove under various conditions of 'normal' driving, economical driving and driving with information encouraging coasting and discouraging harsh acceleration. In general, participants significantly changed their behaviour when asked to drive 'economically', mirroring results found elsewhere in the literature (e.g. Evans 1979; Van der Voort et al. 2001; Waters and Laker 1980), and suggesting that people are already largely aware of how to drive economically. Additional information discouraging excessive accelerations did, however, further reduce these behaviours, a finding in line with those from Birrell et al.'s investigation of vibrotactile information (Birrell et al. 2013). However, it appears that coasting, as a means for fuel conservation, is a tactic that is not as commonly understood or applied. The eco-driving information used in this study had a significant effect on the distance spent coasting, an effect that was not observed when simply asking people to drive economically.

There were large inter-subject differences in many of the variables measured, particularly in the baseline, eco and after trials. For the majority of variables (time taken being a notable exception) the differences were greatly reduced under feedback conditions, implying that the additional information was indeed followed by participants. The extent of this effect was not, however, equal for all participants, nor for all modes of feedback.

For example, the data revealed a decrease in control effort as measured by the same metric used in Azzi et al. (2011) and Mulder et al. (2008), namely throttle pedal position standard deviation. In Azzi et al.'s study, haptic information was shown to have a stronger effect than visual. Though results from the present study did not show *statistically significant* differences between the haptic and visual conditions, the patterns of results for fuel use, mean and maximum throttle position, and brake pedal standard deviation variables (Figures 8.6–8.8 and 8.11) do suggest that visual information alone does not support compliance *across all participants* as much as other stimulus modes do. The analysis of interaction effects suggested that the additional eco-driving information had a stronger influence over those participants with a tendency to drive less economically at baseline. Regardless of whether participants drove more or less economically at baseline, or indeed in the 'eco' trial, all performed at close to the same level under feedback conditions. As aforementioned, however, visual-only information in particular did not have as strong an effect as combinations involving auditory and haptic in those participants that drove less economically at baseline.

These results in combination provide tentative support to the hypothesis that auditory and haptic information foster greater compliance that visual information, but only for those that have a more aggressive driving style. Those with a more economical driving style appear to be influenced equally by all modes of information. One might conclude that should a person be inclined to drive aggressively then they will be more likely to disregard eco-driving information that is less salient, that is, easier to ignore.

Results from the excessive acceleration variable (Figure 8.4; where visual information was significantly less effective than auditory, and not significantly different from the eco trial) adds to the support of this hypothesis, allowing for the tentative conclusion that the visual mode alone is less effective than combinations involving auditory or haptic stimuli (or those stimuli alone). One possible explanation for this is simply that the auditory and haptic stimuli were more salient, and the visual information easier to ignore. This conclusion is supported by comments made during the debriefing session at the end of the experiment (mentioned by 13 of the 30 participants). Although stimulus levels were assessed in pilot studies, and deemed acceptable and noticeable (see Section 7.3.2), the salience of each was not strictly controlled. Given the highly complex interplay of factors influencing stimulus salience across sensory modes (Downar et al. 2002), such controls would have been extremely difficult to objectively implement; however, this is accepted as a limitation of the present study. Indeed, it would be worthwhile to investigate the effect of stimuli of different levels of intensity, or frequency, on driving performance. For example, it is possible that stimuli of certain characteristics could already be associated with particular cues in the environment. For example, a certain auditory tone could already be used as a system warning; such an alert would likely elicit a different type of response than an auditory tone with different characteristics. One could also imagine that the red light presented in the current experiment might bring to mind the appearance of the brake light of a lead vehicle. Such a stimulus would already have been associated with a particular response (i.e. to decelerate, or at least prepare to decelerate). The investigation of these possibilities therefore presents a potentially interesting avenue for further research.

Regarding satisfaction ratings, it was clear that participants did not like the auditory stimulus, whether it was presented alone, or in combination with other stimuli. Although it encouraged compliance (perhaps due through affecting perceived urgency; see Marshall et al. 2007), such low ratings are unacceptable. As research from the medical domain demonstrates (Block et al. 1999), annoyance undermines the effectiveness of any system, as the user will simply ignore it or turn it off.

In terms of perceived usefulness, few differences between stimulus modes were observed. The only observed differences involved either the eco or after trial. The finding that participants rated information in the eco trial as less useful than three of the feedback trials (each of which involved haptic information), but did not rate the baseline trial differently, may be a reflection of the perceived requirement for more information when an extra task is demanded of drivers (i.e. that of driving economically, rather than simply driving as one 'normally' would).

Similarly, the information in the after trial was also rated as significantly less useful than any of the feedback trials. Again, information in this trial was identical to that in the baseline and eco trials (i.e. information from the vehicle's dashboard). One might conclude here that the participants had become accustomed to the information and had begun to rely on it to guide behaviour; upon its removal, participants felt less well supported in the task. This interpretation is supported by comments made by participants in informal debrief interviews; 8 out of 30 made unprompted comments relating to developing reliance on the information.

With regard to workload, this study did not reveal any significant differences between conditions. As such we cannot conclude that visually presented eco-driving information has a stronger effect on workload than does auditory or haptic feedback, a suggestion arising from Wickens' multiple resource theory (e.g. Wickens 2008). In our study, and in Birrell et al.'s (2013), only self-reported, subjective measures of workload were taken (i.e. the NASA-TLX); future research may do well to assess this more objectively, for example via performance on a secondary, distractor task performed concurrently with the driving task.

As was described in the method section, for coasting support the time-to-event at which the information was presented was set at 8 seconds. This was an arbitrary figure considered suitable in pilot studies; it does not represent an ideal distance-to-event for the commencement of the coasting phase for all vehicles or situations. The important issue here, however, is whether or not additional, in-car stimuli can encourage release of the accelerator pedal in order to maximise use of the car's momentum. Results from the present study, in line with those of Hajek et al. (2011), suggest that indeed it can. A potential criticism of this research is that it assumes that it is possible for the vehicle to 'know' what is coming up before the driver does. This has not yet been fully realised on the roads today; however, research from, for example, Muñoz-Organero and Magaña (2013) suggests that this is not a distant reality. They describe an android phone-based system that detects upcoming traffic lights, suggesting to the driver the optimum time at which to take the foot from the accelerator pedal, taking into account rolling resistance and road slope angle (Muñoz-Organero and Magaña 2013). Though the research did not consider the colour of the signal (i.e. green, amber, red), research on traffic-light-to-vehicle communications for fuel conservation (Alsabaan et al. 2013) suggests that this may well be possible.

Furthermore, such developments are not restricted to the academic realm. In Continental's eHorizon project (referred to in Chapter 1) a system is under development that combines data from digital and topographical maps, vehicle sensors and GPS, in order to prepare for upcoming road events (Continental 2015). Although the project focuses on predictive control of vehicle systems, the possibility of using the same information to inform the *human* driver of upcoming events (i.e. by feeding into an in-vehicle information system) is clear to see. Even when this kind of information does become available in production vehicles, however, the question of timing will still remain. Research from Staubach et al. (2014a) goes some way to address this topic, showing that longer lead-times support greater coasting behaviours. Indeed, this topic is the focus of the next chapter.

The choice of the 70% threshold for discouraging excessive acceleration was also an arbitrary number that was considered appropriate in pilot studies. Rather than focus on particular vehicle parameters, this research was interested in the effect on drivers of multisensory stimuli designed to guide certain behaviours. Indeed, to this aim it can be concluded that feedback involving auditory and or vibrotactile stimuli does discourage such behaviours, and that visual stimuli are less likely to do so. The choice of threshold would, however, likely have had an effect on compliance with the alerts, on user acceptance of the system, and on resulting driving performance and fuel use. Additional research would be required to assess *how* these variables would be affected by, for example, an input threshold of 60% or 80%. We would argue here, however, that for the comparison of the various stimulus modes and combinations, the threshold itself is not as important; it was the same across conditions, hence the ability to compare the effects *between* those conditions remains valid.

Continuing with the question of stimulus timings, the actual optimal values for either acceleration or coasting guidance will vary greatly with the vehicle and the context of use. Assessing different levels would therefore present a highly worthwhile avenue for research. Additionally, the present study was based entirely in a driving simulator. Assessing a vibrotactile system on the road, with the presence of ambient vibrations, is also necessary to take such an idea forward. It could be argued that vibrations already naturally present in the moving vehicle (i.e. those arising from tyre/road surface interaction, and from the vehicle's engine itself) might mask additional stimuli. Such a question, and whether certain stimulus frequencies are more distinguishable over ambient vibration than others, requires additional research.

At this point it is worth briefly mentioning the general validity of driving simulator research as a means for investigating driving performance. In terms of the present study, it is important to accept the limitations inherent in a fixed-base simulator when assessing responses to acceleration and deceleration events. Such equipment does not allow the driver to feel the vestibular cues that would be present in a real on-road environment (or, to a lesser extent, in a moving-base simulator). Before an eco-driving system such as that described in this book could be built in to a commercial, on-road vehicle, it would be necessary to extend the current line of research by conducting on-road studies. Additionally, the resolution of the images seen by participants was clearly not as high as would be experienced in the real world. Although all events in this study could be seen *before* presentation of the additional stimuli, it is not possible to state whether or not they would be more easily noticed in an on-road environment (though, admittedly, this would likely depend to a great extent on weather conditions). Again, such questions could only be properly addressed by extending this research with on-road studies using an instrumented vehicle.

Despite these limitations, simulator-based research as a whole still offers a valuable stepping stone from basic, desktop-based research, to on-road trials involving instrumented vehicles. Not only do simulators provide a good compromise between representative design (i.e. an experimental environment that close-matches the real-world environment it attempts to represent) and experimental control (e.g. Jackson and Blackman 1994), but they also elicit driving behaviour that corresponds, to a reasonable degree, to that which would be seen in an instrumented vehicle (e.g. Blaauw 1982; Stanton et al. 2001).

It would also be valuable to compare the effects of this type of vibrotactile information with the effects of force or stiffness feedback, such as that reported in the aforementioned Jamson et al. (2013), Azzi et al. (2011) and Mulder et al. (2008, 2011) articles, both from an objective, quantitative viewpoint and in terms of user acceptance and workload. As described in the introduction to this chapter, the great majority of research on haptic accelerator pedals focuses on force or stiffness feedback; a comparison with vibrotactile is, to our knowledge, completely lacking in the driving literature.

Also of interest is the effect of gear-choice support on driv\. user acceptance. In Staubach et al.'s work (Staubach et al. 2013, 20\ accelerator-based information was successfully employed in the supp\ gear-choice behaviours. One reason for the decision not to include this ty\ back was to allow for generalizability to vehicles with non-conventional dr\ (e.g. electric vehicles). Currently these vehicles represent only a small propo\ of those on the road; however, the number of these vehicles registered each year\ consistently growing (Society of Motor Manufacturers and Traders 2016). Research focusing on these technologies is, therefore, merited. Another reason for omitting gear-change support was to retain the ability to compare information presented at the site of control with information of equal content presented at an incongruous location. This would not have been possible with gear-change support (as discussed in the Introduction to this chapter). Nevertheless, it important to recognise that gear choices do still have a significant effect on fuel-economy, and are indeed worthy of support by an in-vehicle eco-driving support system.

In contrast to early research from Evans (1979), and Birrell et al.'s more recent work (2013), the data presented here suggest that eco-driving benefits are realised at the cost of journey time. Simply slowing down was not the sole cause of the reduction in fuel use; participants did not take significantly more time to complete the 'after' trial compared to baseline, although they did use significantly less fuel. The link between fuel use and journey time, and the effect that additional stimuli have on this link, remains unclear. This is potentially important for the encouragement of eco-driving. Increases in journey time are seen a significant barrier to the uptake of eco-driving more widely (e.g. Harvey et al. 2013); however, as evinced here, significant fuel savings can be achieved without significant journey time increases. Given the time differences between baseline and eco trials it appears that average speed and fuel use are closely linked in the minds of the participants. This conclusion is entirely in line with early work from Waters and Laker (1980), and results from the survey study reported in Chapter 3. The question remains, therefore, of how to manage drivers' expectations. Specifically, how do we display to driver that it is possible to drive significantly more economically with only marginal increases in journey time? Solving this problem would reduce a major barrier to the wide-scale uptake of eco-driving.

In terms of the question regarding the SRK taxonomy and the EID framework, conclusions are not so easily drawn. Although the haptic in-vehicle eco-driving support has, in this book, been justified using the theoretical arguments arising from EID and the SRK taxonomy, and from the DMI approach to interface design (Hutchins et al. 1986), results from this study cannot be used to test the validity of these theoretical justifications.

First, haptic, accelerator-based information was argued to more likely support skill-based behaviour than information in other modes. Second, it was suggested that previously held beliefs and habits would play less of a role in influencing behaviour at this level of cognitive control than when interacting at the rule- or knowledge-based level of cognitive control, where there is a more conscious interpretation of the stimuli–response pairing. It was therefore suggested that more or less pro-environmental individuals would be affected equally under haptic information conditions, but that under auditory and visual conditions (argued to be more likely to guide behaviour at the rule-based level) behaviour would be affected differentially in these two groups.

Results from the environmental attitude inventory cannot be used to confirm this suggestion; there were no differences in driving performance between those with higher or lower scores on the questionnaire, either in baseline driving style or in 'eco' driving style. Although one might expect those more environmentally minded (i.e. more likely to *report* energy-saving behaviours) might drive more economically in general, this was not the case. This reflects results from the survey study presented in Chapter 3; only weak relationships were found between environmental attitudes and self-reported eco-driving behaviours. Results in this chapter suggest that the relationships between environmental attitudes and *observed* eco-driving behaviours are even weaker, if they exist at all. The environmental attitude variable was not, therefore, useful in assessing the potential for haptic information to support behaviour at the skill-based level of cognitive control.

The only between-subjects variable from which significant interaction effects *did* arise was the baseline fuel use distinction, that is, the median split that separated participants into those that used less, and those that used more fuel at baseline. In particular, participants of a more aggressive driving style reacted to visual stimuli, stimuli that have been argued to be more likely to exert control at the rule-based level, to a lesser extent than to haptic information, stimuli that have been argued to be more likely to exert control at the skill-based level. It is possible to argue that this group separation is a reflection of underlying habits (i.e. the tendency to drive more or less efficiently), and that success of haptic information in *all* participants, and the differences in the effect of visual information on the groups, is an indication that the two stimulus modes do indeed guide behaviour at different levels of cognitive control. Haptic information supports skill-based behaviour, thereby bypassing habits, whereas visual information supports rule-based behaviour, therefore habits still influence performance.

However, auditory information also had a strong effect on all participants; this was also hypothesised to act at the rule-based level. Although one could argue for the success of haptic information over visual from an EID- or SRK-informed theoretical standpoint, given the ability of auditory information to foster the same level of compliance one could also argue that these stimuli are simply more salient (as discussed above). Although it is possible to justify the haptic feedback using SRK arguments, it is not possible to justify SRK arguments using the results of this study.

It is important to reiterate that this chapter does not represent the first to discuss haptic interfaces in the driving domain from an EID perspective. As has been previously described, Lee et al. (2004) argue for the use of EID as a guiding theoretical framework in the development of in-car haptic interfaces, arguing that they would be especially suited to supporting skill- and rule-based behaviours. The article states that, for the support of skill-based behaviour, 'haptic signals should have a direct analogical link to the motor response requirements – people should be able to act directly on the displayed information' (Lee et al. 2004, 844). The research presented in this chapter does just this; however, to provide empirically derived evidence attributing the benefits of haptic information to its ability to support the skill-based level of cognitive control, rather than simply to its salience, is something that we cannot do here.

It would be worthwhile to investigate this more closely. For example, it would be interesting to investigate reaction times to stimuli presented at the site of control (e.g. vibrations to foot, responses made with the foot) and stimuli presentations at incongruous locations (e.g. stimuli to the hand, response with the foot), and compare these with incongruous modes as well, for example auditory stimuli and responses made with the foot or hand. This would present a less representative design (i.e. less reflective of the driving ecology; Brunswik 1956) than that used in this study, hence generalisations to the driving domain would be less easily made, but it may help to answer the more fundamental questions surrounding the SRK taxonomy and the ability of haptic information, presented at the site of control, to guide behaviour at the skill-based level.

8.7 CONCLUSIONS

This experimental evaluation of multimodal, in-vehicle, eco-driving support has demonstrated that although participants are generally aware that excessive accelera-tion is to be avoided when eco-driving, they can be encouraged to do so further with the addition of auditory and haptic in-car information. Furthermore, using only the vehicle's momentum (i.e. coasting) when approaching an event necessitating slowing is a technique that is little used when asking participants to drive economically, but can be supported with in-vehicle feedback. Though visually presented information is effective in supporting eco-driving behaviours, given the pattern of results across the different measures, it does not appear to foster the same level of compliance as haptic or auditory information (or combinations containing those modes) across all participants. Auditory information, however, was almost universally disliked.

There is an argument for the ability of haptic information to encourage behaviour at the skill-based level of cognitive control, and this study lends some support to that argument; however, more research is necessary to separate the potential effect of salience (i.e. the haptic and auditory stimuli were not easy to ignore) from the effect of combining action and control surfaces as argued by the philosophies behind DMI (Hutchins et al. 1986), the SRK taxonomy (Rasmussen 1983) and EID (Rasmussen 1983; Rasmussen and Vicente 1989). For the next, and final experimental chapter of this book, however, rather than delve further into theory, a more practical focus was taken. This reflects one of the primary aims of the project described in this book, namely to develop an eco-driving support system that is not only effective in its support of efficient driving behaviours, but is also favourably received by drivers. Without such favourable opinions, the system is likely to either be ignored, or (if possible) turned off.

This chapter has demonstrated that coasting behaviours are worthy of support in the vehicle, and that vibrotactile feedback is a suitable means for conveying infor-mation that supports such behaviour; the following chapter investigates when that information might best be presented, and what happens when that information is removed, without warning, halfway through a driving session.

9 When to Give Those Good Vibrations

9.1 INTRODUCTION

In the previous chapter the first experimental analysis of the system described in Chapter 7 was presented. This chapter presents the second; however, where the previous chapter was concerned with theoretical arguments arising from ecological interface design (EID) and the skills, rules and knowledge (SRK) taxonomy, this chapter focuses solely on practical fuel-use benefits and user acceptance. As such, this chapter presents a direct follow-on from the previous, but with a practical, eco-driving focus, rather than a theory-driven perspective.

To this end two major alterations were made to the system given the results presented in the previous chapter. First, only vibrotactile stimuli were used. This type of information was both effective (in supporting the target behaviours) and well accepted by participants, hence was considered suitable for in-vehicle eco-driving support. Second, only stimuli supporting coasting behaviours were employed; stimuli discouraging excessive acceleration were not investigated in the experiment described in the coming pages.

As was described when discussing results of the previous chapter, participants made a number of comments that suggested that the two stimulus triggers, that is, that the participant had exceeded 70% or that they were approaching an event necessitating deceleration, were, on occasion, confused with one another. Furthermore, although people could be encouraged to do so further, the results suggested that participants were already aware of the reduction of excessive acceleration as an eco-driving strategy. Coasting behaviours, on the other hand, were not exhibited when asking people to drive efficiently; these were only evident upon addition of the eco-driving information (i.e. the stimuli presented 8 seconds before a slowing or stopping event). These behaviours therefore provided the sole focus of this chapter.

Indeed, this aspect of eco-driving, that is, the anticipation of the road ahead in order to maximise the coasting phase of the vehicle ahead of slowing and stopping events (thereby minimising kinetic energy losses), has, in isolation, been shown to have a significant effect on overall fuel economy (Thijssen et al. 2014). For example, the reader may recall the referencing of work by Hajek et al. (2011) at various points in this book. The article describes an investigation of deceleration advice alone (i.e. encouraging enhanced coasting phases), finding 7.5% fuel savings compared to baseline (Hajek et al. 2011). Additionally, reference has been made to a number of studies conducted by Staubach et al. (2012, 2013), in which a system that used both visual and haptic feedback for the support of gear choice and enhanced coasting was investigated. It is, however, their most recently published work that is of particular significance for this chapter, insofar as it is concerned with the investigation of different stimulus timings.

In Staubach et al. (2014a) the distance ahead of the event at which information was presented was specifically investigated, with a two-stage process being tested. In the first stage the information was provided at the time when coasting *in neutral* would have participants come to a stop for a red traffic light. Should the participant have disregarded this advice, a second signal was provided at the point at which coasting *in gear* would have them stop at the correct time (i.e. at a later time point than for coasting in neutral). Though increased coasting phases were supported (compared to baseline) the earlier advice was not well received by participants, the reason for which was put down to the advice coming before the participant could see the reason for that advice, for example, the traffic lights (Staubach et al. 2014a). Similar results were also reported in Staubach et al. (2014b). In this study the increased use of coasting (in conjunction with earlier gear changes) resulted in fuel savings of 15.9% and 18.4% for urban and rural scenarios respectively. The system was generally well received; however, though acceptance of the timing of the coasting advice was not expressly investigated, some participants did state that the advice was presented too early (Staubach 2014b).

In the study described in the previous chapter the timing of coasting advice was held constant at 8 seconds before the slowing event (a figure based on the pilot studies described in Chapter 7). As aforementioned, however, this does not represent an idealised distance for all cars (in terms of engine parameters), nor for all people (in terms of both journey-specific requirements and high-level eco-driving goals). Hence the research presented in this chapter addresses the effect of different stimulus presentation timings.

In line with results from Staubach and colleagues' research (Staubach et al. 2013, 2014a, 2014b) it was hypothesised that the participants would follow the coasting advice inasmuch as the further away the advice was presented, the greater the coasting phase and, consequently, the greater the overall fuel efficiency of the drive. Furthermore, it was hypothesised that participants would have higher acceptance for information presented closer to the event than that which is presented further away. This is again in line with Staubach et al.'s finding concerning participants' acceptance of early advice.

In the debrief interviews conducted at the end of each participant's involvement in the study described in Chapter 8, a number of participants commented on becoming reliant on the system, expressing concerns about times at which it would not be available and times at which the system might not be present, for example if one should move from a vehicle that does have the system to one that does not, or if the system should drop out unexpectedly. A second section to the experiment described in this chapter was therefore included, in which the coasting advice was removed – without warning – halfway through the trial, thereby simulating system dropout. In Chapter 8 there were few differences in driving performance between feedback trials and the final trial in which no information was presented. This final trial, however, was a separate driving session in which participants drove a route that was similar to the previous routes; the current study differs insofar as it assesses the removal of information halfway through a *novel* route, thereby simulating system failure in an unknown environment. No specific hypotheses were made regarding driving performance in this section; the question is left open.

9.2 METHOD

This experimental study used a repeated-measures design, using only within-subject measures. As described above, the study was split into two sections. In the first, a single independent variable (i.e. the timing of eco-driving information) was used. This had four levels; one baseline condition and three conditions in which eco-driving information was presented (at different lead times). The order in which participants experienced the three experimental conditions was randomised and counterbalanced. Section two consisted of one driving trial, performance in the first half of which was compared with performance in the second. In the first half, participants received eco-driving information; in the second half they did not. The single independent variable therefore contained two levels. The procedure is described in more detail below, and summarised in Table 9.1 in Section 9.2.3 below. The two sections shared eight dependent variables, each of which related to an aspect of driving performance. Section 1 used an additional four dependent variables not used in section 2; these measured self-reported workload, and subjective ratings of perceived usefulness of, and satisfaction with the system. Section 2 used an additional three dependent variables not used in section 1; these related to general vehicle control and safety. All independent and dependent variables are summarised in Table 9.2 in Section 9.2.3 below.

9.2.1 Participants

Twenty-four participants (14 male, 10 female), aged between 23 and 60 (M = 34.71, SD = 13.08), were recruited through a convenience sample. All participants had previously participated in (and completed) the study described in the previous chapter. Participants reported annual mileages ranging from 300 to 15,000 miles (M = 6741.67, SD = 4173.09) and all had full EU driving licences (held for between 2 and 42 years, M = 15.25, SD = 12.81), with at least 1 year's experience driving on UK roads. Participation was entirely voluntary, all participants gave fully informed consent and none were paid for their time. Ethical approval was sought from, and granted by, the University of Southampton's Ethics and Research Governance committee.

9.2.2 Apparatus

9.2.2.1 Driving Simulator and In-vehicle Information

Trials were again conducted using the University of Southampton Driving Simulator, described in detail in Chapter 7. The vibrotactile feedback was provided via the same device used in the study described in the previous chapter, that is, an array of six, 3-volt vibrating motors attached to a metal plate, which was in turn attached to the accelerator pedal. The same vibration characteristics were once again employed (at

12,000 rpm, deemed acceptable in pilot studies; see Chapter 7); however, the system described in Chapter 7, and used in Chapter 8, was modified. Whereas 11 settings were required for the previous experiment (reflecting the 11 scenarios; see Chapters 7 and 8), only 2 were required for this experiment (reflecting the use of only 2 scenarios, described below). The information (including the distance down the road at which each event occurred) contained within the box (shown again below, Figure 9.1) was necessary for the presentation of stimuli ahead of each event (see Chapter 7 for a description of the system's working, and the requirement to include scenario information on this device). A second rotary knob was also added; this allowed for the selection of the different stimulus lead times to be assessed. Three timings were chosen for assessment; short (4 seconds ahead of the event), medium (8 seconds ahead of the event), and long (12 seconds ahead of the event). On the device, depicted in Figure 9.1, the lower dial reflects these three timings; close (short lead time), medium (medium lead time) and far (long lead time).

9.2.2.2   Questionnaires

To measure workload the raw NASA-TLX was once again used (Hart 2006; Hart and Staveland 1988). As the reader may remember from the previous chapter, the questionnaire elicits subjective workload ratings, across six items (perceived mental, physical and temporal demands, frustration, effort and performance), each on a 20-point scale. To assess the participants' general acceptance of the system the Van Der Laan acceptance scale was also adopted a second time (Van Der Laan et al. 1997). As before, this elicits responses on two scales (perceptions of usefulness of, and overall satisfaction with, a transport telematics system) across nine items, each on a 5-point scale from −2 to +2.

Further to the questionnaires used in the study described in the previous chapter, the participants for this experiment were also required to respond to an additional

FIGURE 9.1 Scenario and stimulus timing selection control box.

set of questions adapted from Staubach et al. (2014b). These questions assessed system acceptance in terms of usefulness, ease of use, attitude towards and behavioural intention to use the system (see Table 9.4 in the results section, below, for all questions). In addition to deceleration behaviours Staubach et al. also investigated gear changes and acceleration strategies; questions relating to these behaviours were not included in the current research. The remaining 15 questions, worded identically to those used in Staubach et al. (2014), were scored on a 7-point Likert scale (1 = strongly disagree, to 7 = strongly agree), a modification to Staubach et al.'s 5-point scale. An additional item was included in order to assess participants' subjective ratings of the stimulus presentation timings; this item was scored using a 9-point scale (from −4 = information presented too early, to +4 = information presented too late).

9.2.2.3 Driving Scenarios

Two different driving scenarios (i.e. road environments) were developed for this study, both of which simulated a mix of rural and urban driving on UK roads. The first was 16,200 feet (4.93 km) in length, 4630 feet (1.41 km) of which was through a village at speed limit of 30 miles per hour (mph), the remainder of which (11,570 feet/3.53 km) was through rural, country roads at the UK national speed limit (i.e. 60 mph for roads of that type). The second scenario was 31,400 feet (9.57 km) in length. Of this, 9260 feet (2.82 km) was through two town environments with a 30 mph speed limit, the remainder (22,140 feet/6.75 km) being on rural roads with a 60 mph speed limit. Once again, these lengths were chosen as they are comparable to the length (at 5 to 10 minutes) of the most common single trip distance bracket (Department for Transport 2014).

The first scenario contained three traffic light intersections, one of which was in a 60 mph section, the other two being in the 30 mph village section. Traffic lights were timed such that they changed from green to amber when the driver was 10 seconds away from them. The amber light remained on for 4 seconds, and then red for 15 seconds; this ensured that drivers had to slow to a complete or almost complete stop (depending on their braking and coasting behaviours). The scenario also contained two road curvatures, one in a 60 mph section (requiring the driver to slow to approximately 20 mph) and one in the 30 mph section (requiring the driver to slow to approximately 12 mph). Data was captured (at 30 Hz) from 200 feet from the start, until the end of the run.

For data capture the second scenario was split into two halves, each of identical length. Though each half was different in appearance to the other (i.e. buildings, road textures, pedestrians, trees, and road verge materials and gradients were all entirely different), each had an identical topography and each had an identical number of slowing events, at identical spacing. These were identical to those in scenario one; three traffic lights (one of which was in a 60 mph section, the other two being in a 30 mph section), two road curvatures (one in a 60 mph section requiring a cornering speed of 20 mph, the other in a 30 mph section, requiring a cornering speed of 12 mph) and one transition from a 60 to a 30 mph zone. In each of the scenarios the participant had ample time to get back up to the posted speed limit well in advance of the next stopping or slowing event.

9.2.3 PROCEDURE

The study lasted for approximately 1 hour. Each participant was subjected to seven driving trials, six of which had the participant repeatedly driving the first, shorter scenario, the seventh requiring them to drive the second, longer scenario. No data were recorded on the first two trials; these were practice trials to allow the participant to familiarise themselves with both the simulator (though note that all participants already had experience in the simulator) and, more importantly, with the route. The third trial acted as the baseline trial. Participants, although knowledgeable of the eco-driving focus of the study (all had participated in the experiment described in the previous chapter), were not asked to drive economically in this trial, rather they were asked to drive as they normally would. Data were captured but no additional eco-driving advice was provided. Participants completed the Van Der Laan and raw NASA-TLX questionnaires upon completion of the trial.

Trials four to six each required the participant to drive with the addition of vibrotactile feedback designed to encourage coasting when approaching slowing events (i.e. traffic lights, lower speed limits and road curvatures). The information was presented at the moment at which they were required, or suggested, to remove their foot from the accelerator pedal. The stimulus onset timing varied across the three trials, at 4 (short lead time), 8 (medium lead time), or 12 (long lead time) seconds ahead of the event. The reader may recall that a stimulus lead time of 8 seconds was used in the study described in the previous chapter, a figure accepted, based on extensive piloting, as a balance between early action and user acceptance; the figures used here therefore present equidistant steps closer to, and further from the event necessitating action.

The order of the trials was randomised and counterbalanced across participants. Stimuli were provided continuously when, and for as long as, the accelerator pedal was depressed and the vehicle was travelling at a speed greater than that necessitated by the event. Stimulus presentation would stop as soon as the accelerator pedal was released, as soon as the target speed had been reached, or as soon as the event had been passed. For road curvatures this target speed was either 20 mph or 12 mph and for speed limit changes this was 30 mph. For traffic lights this was set at 5 mph rather than 0 mph as there was the possibility for the car not to come to a complete stop. The participants completed the raw NASA-TLX, Van Der Laan scale and the system acceptance and usability scale adapted from Staubach et al. (2014b) following each of these trials.

The same route was used for the first six trials for two primary reasons; to allow for comparability between data sets, and to simulate a known route, for example the daily commute to work (hence the two training trials on the same route). This allowed for the investigation of the effect of haptic feedback in novel situations (i.e. on an unknown route), and the effect of system failure. This was the purpose of trial seven. In the first half of this longer scenario (described above) the participants received the information from the vibrotactile accelerator pedal at 8 seconds before the event, the medium distance value. Information was then turned off for the second

TABLE 9.1

Summary of Experimental Design

Trial Number	Description	Approx. Length	Feedback	Questionnaires	Route
1	Simulator training	4 minutes	None	None	
2	Simulator training	4 minutes	None		
3	Baseline trial	4 minutes	None	NASA-TLX, Van Der Laan	
4	Experimental trial	4 minutes	Vibrotactile at 4, 8 or	NASA-TLX, Van Der	1
5	Experimental trial	4 minutes	12 seconds before	Laan, usability and	
6	Experimental trial	4 minutes	event (randomised)	acceptance questionnaire	
7	Experimental trial	8 minutes	Vibrotactile at 8 seconds before event for first half, none for final half	None	2

half of the route; the participant was not informed of this, neither before nor during the trial. The procedure is summarised in Table 9.1.

The independent variable for the first part of the experiment (trials one to six) was stimulus timing and consisted of four levels, reflected in trials three to six in Table 9.1. The independent variable assessed in the second part of the experiment (trial seven) had two levels; with information and without information. Statistical analyses of these two sections were entirely separate. These variables are summarised in Table 9.2, below. Table 9.2 also summarises all the dependent variables used in the study.

9.3 RESULTS

Results presented below are split into two sections, with part one being the investigation of the effect of the three different stimulus timings on coasting behaviours (trials one to six in Table 9.1), and part two being the investigation of the effect of removing coasting advice, without warning, halfway through a novel route (trial seven in Table 9.1).

9.3.1 PART ONE: LEAD TIME MANIPULATION

As all variables met the necessary conditions for use of parametric statistical analyses (see Field 2009), a MANOVA was performed to test for differences between conditions (baseline, 4, 8 and 12 seconds lead time) for the variables outlined in Table 9.2. This resulted in statistical significance: $F_{(24, 180.42)} = 6.009$, $p < 0.0005$; Wilk's $\Lambda = 0.182$, partial $\eta^2 = 0.43$.

Subsequent univariate tests revealed significant differences between groups for all the variables measured; these are presented in Table 9.3. Mauchly's test of sphericity

TABLE 9.2
Summary of Independent and Dependent Variables

Variable Type	Variable Name	Variable Description
IVs	Condition (Part 1 of the study)	Experimental condition in part 1 of the experiment, with four levels; baseline, short lead time, medium lead time, long lead time
	Section (Part 2 of the study)	Section of part 2 of the experiment, with two levels; first half of the route, with coasting support, compared to second half of the route, without coasting support
DVs	Time taken	Time taken to complete each trial (in seconds)
	Fuel use[a]	Fuel used in each trial (metric calculated by the simulator software[a])
	Throttle mean	Mean throttle position (0 to 1) across trial
	Throttle SD	Throttle position (0 to 1) standard deviation across trial
	Brake SD	Standard deviation of brake pedal input force (in pounds)
	Distance coasting	Total distance spent travelling forward (i.e. >0 mph) without depressing throttle (in feet)
	Brake use	A measure of overall brake use, calculated by taking the total area under the curve created by brake pedal input force by time across entire trial
	Accelerator use	A measure of overall accelerator use, calculated by taking the total area under the curve created by accelerator pedal input force by time across entire trial
	Road edge excursions	Discrete number of times the vehicle goes beyond the left-hand road boundary (used only in part 2 of the study)
	Centre line crossings	Discrete number of times the vehicle crossed the road's centre line, into the adjacent lane of oncoming traffic (used only in part 2 of the study)
	Lane position standard deviation	Standard deviation of the vehicle's lateral position (in feet), an indicator of driving control performance (used only in part 2 of the study)
	Raw NASA-TLX	Raw NASA task load index workload score (used only in part 1 of the study)
	Satisfy	Satisfaction score on the Van Der Laan acceptance scale (used only in part 1 of the study)
	Useful	Usefulness scores on the Van Der Laan acceptance scale (used only in part 1 of the study)
	System acceptance and usability scales, from Staubach et al. (2014)	Questionnaires providing additional measures of perceived system usefulness, ease of use and intention to use, and a measure of the perceived appropriateness of stimulus timing (used only in part 1 of the study)

[a] Fuel use was calculated by the STISim software using the following formula: Total fuel used = Total fuel used + DT × (Specific fuel consumption × (9.55 × Torque × Engine speed (in revolutions per minute) / 5252)), where DT = frame rate (60 Hz) and specific fuel consumption = 0.278 lb/bhp-hr.

TABLE 9.3
Post-MANOVA Univariate Test Results

Dependent Variable	F (df)	p Value	Partial η²
Time taken (seconds)	12.176 (2.455, 56.471)	<0.0005	0.346
Fuel use (metric calculated by simulator software; see notes to Table 9.2)	8.538 (2.175, 50.032)	<0.0005	0.271
Throttle mean (0 to 1)	33.477 (3, 69)	0.027	0.593
Throttle SD (0 to 1)	5.465 (3, 69)	0.002	0.192
Brake use (area under the curve created by brake input (in pounds) by time)	6.412 (3.258, 40.678)	0.005	0.218
Brake SD (in pounds)	7.560 (3, 69)	<0.0005	0.247
Accelerator use (area under the curve created by throttle input (0 to 1) by time)	33.892 (3, 69)	<0.0005	0.596
Distance coasting (in feet)	61.999 (3, 69)	<0.0005	0.729

was violated for the time taken, fuel use and brake use metrics, therefore results for these metrics are reported in terms of the Greenhouse-Geisser correction. All other metrics satisfied the conditions of sphericity and are therefore reported as such.

Pairwise comparisons were conducted for all the variables listed in Table 9.3, with the Bonferroni correction applied, revealing a wide variety of significant differences between the conditions. Perhaps not surprisingly (especially given results presented in the previous chapter) the long lead time condition took significantly longer to complete (at M = 344.5 seconds, SD = 30.08) than baseline (M = 327.8, SD = 38.83; 5.1% difference, p = 0.003), short (M = 320.3, SD = 26.13; 7.6% difference, p <0.0005) and medium (M = 329.6, SD = 30.42; 4.5% difference, p <0.0005) lead time conditions. No other conditions differed significantly from each other for this variable.

In terms of fuel use, none of the baseline, short lead time or medium lead time conditions differed significantly from one another. In contrast, in the long lead time condition significantly less fuel was used (at M = 0.102, SD = 0.008; see Table 9.2 for metric calculation) than in the baseline (M = 0.115, SD = 0.014; 11.3% difference, p = 0.001), short lead time (M = 0.113, SD = 0.015; 9.7% difference, p = 0.009) and medium lead time conditions (M = 0.110, SD = 0.010; 7.3% difference, p = 0.001).

In terms of mean throttle position (measured from 0 to 1), there were no significant differences between baseline and the short lead time condition. There were, however, differences between all other conditions. Mean throttle position was significantly lower in the medium lead time condition (at M = 0.421, SD = 0.033) than either baseline (M = 0.446, SD = 0.053; p = 0.009) or short lead time (M = 0.447, SD = 0.042; p = 0.017), and in the long lead time condition mean throttle position was lower again (at M = 0.376, SD = 0.025). The differences between the long lead time condition and all other conditions was significant at the p <0.0005 level.

There was only one significant pairwise difference for the standard deviation of throttle position, between baseline (M = 0.289, SD = 0.049) and the long lead time condition (M = 0.310, SD = 0.043; p = 0.009). Perhaps surprisingly it was the baseline condition that saw lower throttle position standard deviations. The case is similar for brake pedal standard deviation (the depression force, measured in pounds); for this metric, the only two significant comparisons were between the long lead time condition (M = 5.20, SD = 3.61) and baseline (M = 7.25, SD = 5.16; $p = 0.005$), and between the long and short lead time conditions (M = 7.78, SD = 4.97; $p = 0.002$). In contrast to throttle SD, for both cases brake pedal SD was lower in the long lead time condition.

For the brake use variable (the area under the curve created by brake input force by time) there were no significant differences between the first three conditions (baseline, and short and medium lead time); however, the long lead time condition, at M = 292.3, SD = 280.7, differed significantly from all three; the brake was used significantly less than in the medium (M = 458.2, SD = 446.3; $p = 0.024$) and short lead time conditions (M = 519.8, SD = 424.9; $p = 0.002$), and less than in the baseline condition (M = 525.7, SD = 487.1; $p = 0.006$).

The amount of accelerator pedal usage and the distance spent coasting both showed a number of patterns across the four conditions, each containing a variety of significant pairwise differences. Results for these variables are presented in Figures 9.2 and 9.3.

Due to the non-parametric nature of the data, results from the Van Der Laan acceptance scale were analysed using the Friedman test. For satisfaction scores a significant effect of stimulus lead time was found; $(X^2_{(3)} = 13.622, p = 0.003)$.

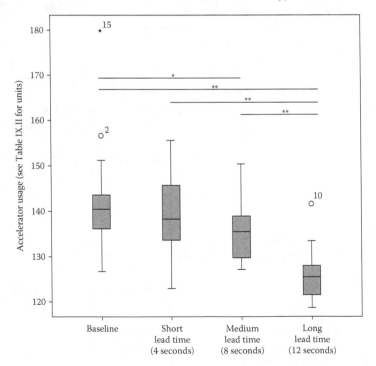

FIGURE 9.2 Total accelerator usage (the area under the curve of throttle position, 0 to 1, by time, across the whole trial), by condition. Solid lines indicate all significant pairwise comparisons, * $p <0.01$, ** $p <0.0005$.

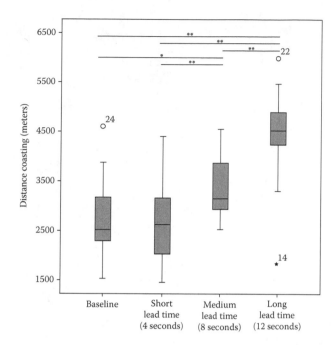

FIGURE 9.3 Distance spent coasting (i.e. travelling forward with zero throttle depression), by condition. Solid lines indicate all significant pairwise comparisons, * p <0.005, ** p <0.0005.

A significant effect of stimulus lead time on usefulness ratings was also found ($X^2_{(3)}$ = 16.850, p = 0.001).

Pairwise comparisons, using the Wilcoxon signed-rank test with Bonferroni corrections applied, were then performed to investigate intergroup differences. For satisfaction scores (measured from −2 to +2) the only significant finding was that the short lead time condition was considered significantly *less* satisfying (at M = 0.052, SD = 0.793) than the baseline condition (M = 0.646, SD = 0.612; Z = −3.019, p = 0.003). Pairwise comparisons revealed more differences for usefulness ratings; these are presented in Figure 9.4.

Friedman tests found no significant effect of stimulus lead time on the number of road edge excursions ($X^2_{(3)}$ = 1.053, p = 0.788), nor on the number of centre line crossings ($X^2_{(3)}$ = 2.033, p = 0.566). There were also no significant differences in NASA-RTLX scores ($X^2_{(3)}$ = 2.962, p = 0.397).

Finally, a summary of results for the questionnaire derived from Staubach et al. (2014b) is presented in Table 9.4.

For the question regarding the timing of the stimulus, it can be seen that the short lead time condition attracted ratings that were *furthest* from zero (with zero being 'at the right time'). The medium and long lead time conditions attracted ratings of similar magnitudes away from 0, though in opposite directions; the former was rated as coming too late, the latter too early. Three one-sample Wilcoxon signed-rank tests were performed (with Bonferroni correction applied, resulting in an alpha value of 0.0167) to assess whether results were significantly different from 0; both short and

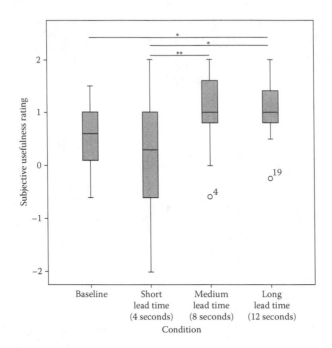

FIGURE 9.4 Van Der Laan usefulness ratings, by condition. Solid lines indicate all significant pairwise comparisons, * p <0.05, ** p <0.005.

long lead time results were ($Z = 3.978$, p <0.0005 and $Z = -2.503$, $p = 0.012$, respectively), medium lead time results were not ($Z = 2.277$, $p = 0.023$).

With regard to the remaining questions, it was found that each of the three scales presented in Table 9.4, namely perceived ease of use, perceived usefulness and behavioural intention to use (as used by Staubach and colleagues), all achieved high internal reliability scores (as evidenced by the presented alpha values); each is therefore treated as a single scale. Results for each scale were summed to provide a single measure, and then subjected to a MANOVA to test for differences between conditions; this resulted in statistical significance, $F_{(6,88)} = 5.839$, p <0.0005; Wilk's $\Lambda = 0.512$, partial $\eta^2 = 0.285$. Subsequent univariate tests were therefore performed, the results of which are presented in Table 9.5 (note that 'perceived ease of use' results violated assumptions of sphericity, hence degrees of freedom are presented in terms of the Greenhouse-Geisser correction for this variable).

Pairwise comparisons, performed using the Bonferroni correction, revealed no significant differences between groups for the perceived ease of use variable. Differences were found, however, for both perceived usefulness and behavioural intention to use. As the usefulness scale had six items, the global score had a possible range of 6 (low usefulness) to 42 (high usefulness). The information in the short lead time condition was rated as significantly less useful (at M = 21.87, SD = 7.20) than in either the medium (M = 29.13, SD = 5.41; p <0.0005) or long (M = 30.96, SD = 4.38; p <0.0005) lead time conditions (with no significant difference between the two latter conditions). For behavioural intention to use,

TABLE 9.4

Means, Standard Deviations and Reliability Coefficients for Answers to the Questionnaire Derived from Staubach et al. (2014b)

Factors	Short Lead Time		Medium Lead Time		Long Lead Time	
	Mean	SD	Mean	SD	Mean	SD
Satisfaction With Information Timing						
Information was presented: too early (−4), just right (0), too late (+4)	2.33	1.31	0.875	1.68	−0.958	1.57
Perceived Ease of Use						
Using the system distracted me from driving (r)	3.17	1.76	2.38	1.24	2.50	1.44
My interaction with the system was clear and understandable	4.50	1.82	5.42	1.28	5.67	0.963
It was easy to follow the information provided by the system	4.71	1.57	5.58	1.25	5.83	1.09
Interacting with the system was frustrating (r)	3.79	1.64	3.25	1.48	3.67	1.63
Interacting with the system was comfortable	4.46	1.56	4.71	1.55	4.63	1.41
Overall, I found the system easy to use	4.92	1.50	5.67	1.24	5.63	1.12
α Coefficients	0.923		0.825		0.834	
Perceived Usefulness						
Using the system increased my awareness of economical driving	3.46	1.67	5.13	1.39	5.67	0.868
Using the system restricted my freedom while driving (r)	3.38	1.38	3.63	1.44	4.33	1.69
Using the system helped me to decelerate in a more fuel-efficient way	3.13	1.73	4.96	1.08	5.71	0.999
Using the system helped me to improve my driving	3.42	1.67	4.75	1.11	5.04	0.954
Using the system would help me to save fuel	3.50	1.64	4.92	1.25	5.63	0.770
Overall, I found the system useful	3.75	1.87	5.00	1.06	5.25	0.989
α Coefficients	0.815		0.826		0.757	
Behavioural Intention to Use						
I believe that the system can help to reduce fuel consumption and therefore CO_2 emissions	3.83	1.58	5.13	1.19	5.67	.868
If I had such a system, I would use it frequently during my trips	3.54	1.56	4.46	1.44	4.58	1.50
I would be willing to pay more for using the system than what I would save in fuel costs and emissions	2.67	1.37	3.13	1.72	3.08	1.59
α Coefficients	0.740		0.869		0.803	

Note: All questions other than that for information timing were given on a 7-point Likert scale; α refers to raw Cronbach's alpha; (r) indicates a reverse-scored item; SD = standard deviation.

TABLE 9.5

Post-MANOVA Univariate Test Results for the

Questionnaire Derived from Staubach et al. (2014)

Dependent Variable	F (df)	p Value	Partial η^2
Perceived ease of use	4.030 (1.348, 31.011)	0.042	0.149
Perceived usefulness	19.555 (2, 46)	<0.0005	0.460
Behavioural intention to use	8.848 (2, 46)	0.001	0.278

summing the three items resulted in a possible range of 3 (low intention to use) to 21 (high intention to use). Participants indicated significantly lower behavioural intention to use the short lead time information (M = 10.04, SD = 3.67) compared to either medium (M = 12.71, SD = 3.92; p = 0.007) or long (M = 13.33, SD = 3.45; p = 0.008) lead time information (again with no significant difference between the two latter groups).

9.3.2 PART TWO: WITH AND WITHOUT VIBRATIONS

As aforementioned, scenario two was split into two halves of equal length, in which there were the same number and nature of events. In the first half participants received vibrotactile stimuli 8 seconds before a slowing or stopping event, in the second half they did not. Performance was compared between these two sections. The same objective measures presented in Table 9.2 were used in the analysis, this time using Hotelling's trace statistic (the appropriate statistic for multivariate analyses involving an independent variable with only two levels; see Hotelling 1931). The MANOVA revealed significant differences in performance between the two sections; $F_{(8, 16)}$ = 3.340, p = 0.019, T^2 = 1.670, partial η^2 = 0.625. Subsequent pairwise comparisons (again, with the Bonferroni correction applied) revealed significant differences between sections for a number of the variables; means and standard deviations for all variables are presented in Table 9.6, below.

For the time taken variable, results were significantly higher in section 1 than in section 2; however, the opposite trend was observed for fuel use, mean throttle position, throttle position standard deviation, brake pedal position standard deviation and total accelerator use. Distance spent coasting was lower in section 2, and total brake use was higher in section 2, though neither result was statistically significant.

A Wilcoxon signed-rank test was used to assess differences between groups for both number of road edge excursions and for centre line crossings as these data did not meet the conditions necessary for performing parametric statistical analyses (see Field 2009). The first half of the route (section 1, with information) saw significantly fewer road edge excursions (Z = 3.153, p = 0.002), but significantly more centre line crossings (Z = 3.938, p <0.0005) than section 2 (without information).

Finally, as data were normally distributed, a paired-samples t-test for lane position standard deviation (measured in feet, and indicative of vehicle control performance) was performed. This revealed no significant differences between performance in section 1 (M = 1.07, SD = 0.347) and section 2 (M = 1.93, SD = 3.48; $t_{(23)}$ = −1.217, p = 0.236).

TABLE 9.6

Means and Standard Deviations for All Measures Included in the Manova Used to Test for Differences Between Driving Performance in the First and Second Halves of Scenario 7 (Section 1, with Additional Coasting Information, Compared to Section 2, Without)

Variable	Section 1 Mean (SD)	Section 2 Mean (SD)	p Value
Time taken (seconds)	301.4 (25.80)	281.2 (31.43)	<0.001
Fuel use (metric calculated by simulator software; see Table 9.2)	0.081 (0.009)	0.089 (0.010)	= 0.001
Throttle mean (0 to 1)	0.363 (0.029)	0.398 (0.052)	= 0.002
Throttle SD (0 to 1)	0.286 (0.042)	0.304 (0.043)	= 0.002
Brake use (area under the curve created by brake input (in pounds) by time)	314.5 (209.7)	425.6 (315.9)	= 0.100
Brake SD (in pounds)	5.82 (3.47)	7.78 (4.81)	= 0.036
Accelerator use (area under the curve created by throttle input (0 to 1) by time)	109.7 (5.66)	112.6 (6.77)	= 0.048
Distance coasting (in feet)	3229.7 (618.5)	2936.1 (859.8)	= 0.111

9.4 DISCUSSION

In the previous chapter the conclusion was made that coasting for fuel efficiency (i.e. maximising usage of the momentum of the vehicle, minimising kinetic energy losses, thereby minimising kinetic energy losses and, as a result, overall fuel consumption) is a strategy particularly suitable for encouragement by an in-vehicle eco-driving support device. Results from the analysis presented above reinforce this conclusion; in general, when provided with vibrotactile stimuli suggesting the removal of the foot from the accelerator pedal ahead of slowing or stopping events, eco-driving performance improves (in agreement with Hajek et al. (2011) and Staubach et al. (2014b)). For most measures this was true for both the medium and long lead time conditions; for the fuel use metric, however, only the longest lead time condition saw significantly lower figures. Results also suggest that when the stimulus is provided too close to the event to which it refers eco-driving performance may even decrease. Looking at Figures 9.2 and 9.3, one can see a greater spread of results for the short lead time condition than for baseline. This pattern can also be seen in Figure 9.4, presenting results of the usefulness section of the Van Der Laan acceptance scale. Although the lack of statistical significance prohibits the making of definitive conclusions, it appears as though some participants waited for the information to be presented before removing their foot from the accelerator pedal on approach to an event, where before (at baseline) they had acted earlier, of their own accord. Informal debriefing conversations with the participants support this. Also, the only significant difference in satisfaction scores (in the Van Der Laan scale) was between the short lead time and baseline conditions, with the short lead time considered significantly *less* satisfying than having no information whatsoever. Furthermore, results from the questionnaire derived from

Staubach et al. (2014b) indicated that participants rated the short lead time stimulus as significantly less easy to use, and would have significantly lower behavioural intention to use it. We therefore conclude that stimulus lead times of this length, that is, 4 seconds, are *not* advisable.

The results presented above suggest that a greater distance-to-event results in greater gains in terms of fuel efficiency, accelerator pedal usage reductions and in the distance spent coasting, thereby supporting one of the hypotheses presented in the introduction. The hypothesis regarding user acceptance, however, is only partially supported. As aforementioned, the shortest distance-to-event was *not* well received by participants; there were, however, only minimal, non-significant differences between the medium and long lead time trials, at least in terms of perceptions of usefulness and the behavioural intention to use. While it is true that ratings for the suitability of stimulus timings resulted in the long lead time stimulus attracting a rating significantly lower than 0 (i.e. the information was considered as coming early), the effect was not a large one (at an average rating of −0.958 on a scale that goes down to −4). Indeed, 6 of the 24 participants rated the information as coming at 'the right time' in this condition (i.e. a score of 0), and 3 rated it as coming too close to the event (scores above 0; see Table 9.7). These result suggest that a stimulus presentation timing of between 8 and 12 seconds ahead of an event, that is, around 10 seconds, may well be optimum, at least in terms of user acceptance.

A possible explanation of the discrepancy between these results and those of Staubach et al. (2014b) is that the participants in the current study could always see the event to which the eco-driving information referred, even in the long lead time condition. Participants in Staubach et al.'s work reported low acceptance of the system when advice was presented very early; in some of these cases, the stimulus came before participants could see its referent (Staubach et al. 2014b). It may be that they simply did not like receiving information when they could not see why that information was being presented; there was no visible event in the scenario with which they could associate said stimulus. Contrast this finding, however, with that of Hajek et al. (2011); participants in this research found coasting advice particularly useful when it was presented far in advance, when the situation to which it applied was not yet visible in the driving scene (Hajek et al. 2011). High acceptance of the advance warning was, however, explained more in terms of safety rather than fuel efficiency (i.e. to help the driver avoid extreme decelerations in rare situations, such as construction sites on rural roads and motorway traffic jams).

TABLE 9.7

Observed Frequencies of Responses to Question Regarding Appropriateness of Stimulus Timing

	Too Close to Event	At the Right Time	Too Far From Event
Short lead time (4 sec)	20	4	0
Medium lead time (8 sec)	13	5	6
Long lead time (12 sec)	3	6	15

In terms of workload, results from the current study can be compared with those of Birrell et al. (2013). Although investigating different eco-driving behaviours (i.e. excessive acceleration discouragement compared to coasting encouragement), both make use of vibrotactile information presented through the accelerator pedal (rather than force or stiffness feedback, used in the great majority of haptic eco-driving research). In both our results and in those of Birrell and colleagues no significant differences in subjective workload ratings between driving normally and driving under feedback conditions were found. Additionally, however, Birrell et al. compared workload ratings under the feedback condition with those from a condition in which participants were simply asked to drive efficiently, without additional in-vehicle support. Here, a significant difference was indeed found, with the feedback condition being rated as incurring lower workload than the efficient driving condition. In the current study no such trial was included that simply asked participants to drive economically, hence we cannot comment on possible effects here; however, in the research described in the previous chapter such a comparison was made, finding no such differences. We would argue, therefore, that the question of whether vibrotactile feedback presented through the accelerator pedal affects workload, subjective or otherwise, remains an open one.

Finally, regarding the open question about the effect of removing information halfway through a novel route, it is difficult to make definitive conclusions, at least in terms of safety and the possibility for the development of reliance. As described in the introduction to this chapter, the inclusion of this second part to the experiment was originally motivated by comments made in the debrief interviews conducted at the end of each participant's involvement in the experiment described in Chapter 8. In particular, a number of participants expressed concern that should the system fail, or drop out, their driving performance could be negatively affected. For example, one could make the argument that if an individual does indeed build up reliance upon a stimulus informing him or her of the appropriate time to remove their foot from the accelerator pedal, when that stimulus is no longer present they may act dangerously late to a given road event. In a road curvature situation, for example, the driver may wait for the in-vehicle system to indicate to them when would be the best time to remove their foot from the accelerator pedal (a stimulus they would have learned to expect). If this were to be the case, reaction to the upcoming event would actually come later than would have occurred if the individual had not previously been exposed to the in-vehicle, coasting support information.

Indeed, results to the first part of the experiment lend some weight to this idea. Driving performance under the short lead time condition was more variable than in the baseline condition, with data suggesting that some participants had reacted later to upcoming events. Informal discussions at the end of each participant's involvement revealed just this; some participants stated that they were waiting for the system to provide them with an indication of the appropriate time to remove the foot from the accelerator pedal, with their resulting action coming *later* than occurred at baseline, where they were aware that no additional information was to be presented.

The rationale behind investigating this effect was not to measure eco-driving performance, but safety. As such, the variables assessed were indicators of driving control performance (which might be expected to decrease if participants act dangerously late to road events), namely the number of road edge excursions and centre line crossings, and the lateral lane position standard deviation (a measure of lateral vehicle control). There was, however, no clear pattern in results for road edge excursions and centre line crossings; results for these two variables differed in opposite directions. These two results taken together do not allow for a simple explanation. There was, however, no difference in the standard deviation of lateral position between the two sections, an additional measure of driving safety. Based on this result we tentatively conclude that even if reliance on the system does develop, this does not necessarily mean that subsequent performance (in terms of safety) without such a system will be deteriorated. This issue, however, clearly requires further attention (e.g. in longitudinal studies).

Regarding the other measures taken in section 2 of the experiment (i.e. those indicative of eco-driving performance), results are in slight contradiction to those obtained in the experiment described in Chapter 8. In part 2 of the study described in this chapter many of the measures of fuel efficiency were seen to change significantly after removal of the information (i.e. efficiency was lower without information compared to when information was present). Although it is not possible to assess whether driving performance after removal of the additional information returned to that which would be seen under conditions of 'normal' driving (the differences between the two scenarios used in the baseline trial and the final trial render the comparison invalid), it is clear that eco-driving performance decreased significantly between the two sections of the final trial. This was not the case in the experiment reported in Chapter 8; here, the majority of results in the after trial (in which no information was provided) did *not* differ significantly from those recorded in conditions with additional, in-vehicle information. A possible reason for this, is that in Chapter 8 the large number of experimental conditions decreased the ease with which statistical significance was reached (i.e. the large number of necessary corrections for multiple comparisons reduced the acceptable significance level considerably). There is, therefore, the possibility that performance does in fact change after removal of additional information. In other words, when additional information is not provided, people do not perform eco-driving behaviours. Deeper investigations of the possible learning effects brought about by such information, and the extent to which people continue to drive efficiently after having experienced in-vehicle, eco-driving advice, certainly requires additional research. Again, longitudinal studies would be aptly suited to address this question.

9.5 CONCLUSIONS

Results from this experimental analysis of a vibrotactile, in-vehicle eco-driving support device show mean fuel reductions of around 11% when advice suggesting removal of the foot from the accelerator pedal is presented 12 seconds ahead of a slowing or stopping event. Of the three stimulus timings investigated, the shortest distance-to-event stimulus trial resulted in the worst performance. Its negative effect on objective measures of performance (even compared to baseline) was also reflected in subjective measures of acceptance; it not well received by participants. The medium and long lead time stimuli, on the other hand, both received high ease of use and behavioural intention to use scores, in addition to their successful support of eco-driving behaviours. Though longitudinal studies are necessary in order to understand potential reliance, system failure and safety effects, the results suggest that vibrotactile eco-driving support presented through the accelerator pedal, particularly for the encouragement of coasting, is a promising avenue for the mitigation of private transport-induced climate change, or indeed any for any road transport-induced climate change (e.g. that caused by freight, taxis or buses).

10 Conclusions and Future Work

10.1 INTRODUCTION

As was described at the very start of the introductory chapter, the work undertaken and described in this book was driven, to a great extent, by two main factors; an interest in ecological interface design (EID) and the skills, rules and knowledge (SRK) taxonomy of human behaviour, and a belief in the society-wide need to reduce our consumption of energy and resources. These two driving forces led to the development of an in-vehicle eco-driving support tool. The design was guided by, and justified using principles from EID and the SRK taxonomy, and ultimately resulted in a system that aimed to encourage coasting behaviours when approaching events requiring the driver to slow down or stop. This was achieved through the use of vibrotactile stimuli presented through the accelerator pedal.

Although the path taken to get to the stage at which an in-vehicle system could be tested and refined was rather winding, involving a variety of methods, a great deal of literature reviewing and a not-inconsequential amount of theorising, the initial question was a relatively simple one; how do we encourage and support the uptake, and fuel-efficient use of low-carbon vehicles? The first notable departure from this question was that low-carbon vehicles did not provide the focus of the work presented in Chapters 3 to 9 of this book. Although this was the starting point, the system that was ultimately developed was one that could be used in any road vehicle, including, but certainly not limited to, low-carbon vehicles. The second main difference between this original research question and the subsequent research effort was with regard to the first part of the question, that is, encouraging the uptake of low-carbon vehicles; the research presented in this book does not address this issue. As results from the survey study presented in Chapter 3 revealed, the relationships between environmental attitudes, eco-driving knowledge and behaviour are highly complex. Although Chapter 3 did not address vehicle-purchasing behaviour specifically, one can be fairly confident in saying that it is impossible to single out any one intervention alone that could stimulate wide-scale behavioural change.

Indeed, the topic of attitudes and behaviour is a vast one, and is one that can be approached from a whole range of perspectives, from social-cognitive theory more generally (see, e.g. seminal work by Bandura 1986), to sustainable consumption more specifically (see, e.g. Jackson 2005 for a review). Suffice it to say here that to motivate people to choose low-carbon vehicles, for example hybrids and full-electrics, as an alternative to conventional internal combustion engine vehicles will require a whole raft of measures, from educational, to political, to financial.

The focus, therefore, was largely on the second part of the aforementioned question, although in a manner more general to all road vehicles, that is, how to encourage efficient use of the vehicle.

One of the main barriers identified in Chapter 2 was the issue of range; given the current state of technology, battery electric vehicles are simply not able to travel as far between refuelling events as are conventional petrol or diesel vehicles. Although technology is improving, and costs reducing (particularly in battery technology, e.g. Nykvist and Nilsson 2015), the barrier remains. This provided the starting point to the remainder of the research effort; when resources are limited, those that are available must be used efficiently, hence the identification of eco-driving as a worthy avenue of exploration. This concept, alongside research from Birrell, Young and colleagues investigating the potential for an EID-inspired interface to support efficient driving behaviours (Birrell and Young 2011; Birrell et al. 2014a; Young and Birrell 2012), led to the focus of the majority of this book; the development, design, testing and refinement of an in-vehicle information system for the support of eco-driving behaviour.

The third significant change in the direction of the research as envisaged at the outset of the project was that EID was not used in its entirety. This is partly to do with the point made above concerning the shift from supporting interface design in low-carbon vehicles, to encouraging the performance of specific eco-driving behaviours in any vehicle. The theoretical journey taken by the research presented in this book, described primarily in Chapters 5 and 6, resulted in an SRK focus, with attention paid to the first of EID's three core principles; to encourage behaviour at the skill-based level of cognitive control by supporting interaction via time-space signals and allowing the user to act directly on the display. Hence the consideration of haptic information presented through the accelerator pedal.

The research effort as a whole has a number of implications in terms of theory, methodology and practice, and although these are all inter-related, each will be discussed in turn.

10.2 THEORETICAL DEVELOPMENTS

This book has discussed the argument that skill-based behaviour occurs largely outside of conscious awareness, (see also Rasmussen 1983, 259); it is characterised by bottom-up processing, relying directly on stimuli in the environment (or system) to guide behaviour. This type of behaviour may therefore be expected to rely less on pre-existing beliefs and habits than would rule- or knowledge-based behaviour, behaviours that can be described as top-down processing.

The decision ladder analysis presented in Chapter 6 went some way to represent this concept; for the expert eco-driver, the timing of the removal of the foot from the accelerator pedal is not a decision that requires careful consideration. It is not one that involves the top part of the decision ladder, that which is concerned with goals and motivations. Rather, in the expert, a short cut is taken from the 'alert' stage, that is, the recognition of an upcoming event necessitating action, to the 'procedure' stage, that is, removal of the foot from the accelerator pedal. The expert does this automatically based on environmental stimuli (stimuli that they will have already amalgamated into a cue for a specific action); however, the suggestion was made that it should be possible to support this in the novice eco-driver with a time-space signal presented by an in-vehicle information system. Importantly, should the signal support behaviour at the skill-based level of cognitive control, the participant's response would not include (in the decision-making process) considerations of goals and options (as presented in the top section of the decision ladder). This can be related back to the aforementioned, unanswered question of encouraging the uptake of fuel-efficient behaviours. Such a system, if it were to support behaviour in this way, should affect all participants equally, regardless of their opinions on fuel use or the environment, or of their motivations. It is not, therefore, a question of motivating people to care about their energy consumption or emissions volumes, but about encouraging eco-driving behaviours even in those who *do not* care. In other words, can we encourage eco-driving behaviours even in those people who do not have pro-environmental attitudes?

The line of reasoning developed from SRK taxonomy principles was expanded upon with regard to the idea of combining action and observation surfaces, a concept featuring in both the direct manipulation interfaces approach (DMI; Hutchins et al. 1986) and in EID itself. Vibrotactile information, presented at the site of control (i.e. the accelerator pedal), was argued to be more likely to exert influence at the skill-based level of cognitive control than would visual or auditory stimuli. The vibrotactile stimulus was argued to be a time-space *signal* (Rasmussen 1983) inasmuch as it presented to the user a physical barrier (surmountable but noticeable) to further pedal depression; it provides a perceptual indicator of a desired action. Auditory and visual presentation, on the other hand, were described more as *signs*, interpreted at the rule-based level of cognitive control (Rasmussen 1983). The stimuli themselves are arbitrary, presenting no physical barrier to further depression of the accelerator pedal (the behaviour in question); there is no combination of action and observation surfaces (as there is with vibrational stimuli), hence do not allow the user to act directly on the area from which information is received (Rasmussen and Vicente 1989).

Furthermore, this concept (i.e. the combination of action and observation surfaces) was linked to the notion of presenting the system's constraints to the driver, a core principle of EID. At the end of Chapter 6 a discussion was offered on the ability of vibrotactile information to display the environmental constraints of a system in a manner that facilitates direct perception. The physical barrier described above is exactly that; it is the constraint inherent to the eco-driving system. Such a stimulus provides a directly perceptible, physical indicator of the eco-driving equivalent of the 'field of safe travel' described by Gibson and Crooks (1938) and discussed by Birrell and Young in their investigations of an EID-inspired in-car interface (Birrell and Young 2011; Young and Birrell 2012). Indeed, we could say that it is the 'field of economical travel' that is being displayed; the stimulus 'displays' an indication of the barrier between economical and inefficient driving. An argument was made suggesting that haptic information, presented through the accelerator pedal (thereby providing a link between action and observation surfaces), would be especially suited to providing a direct, analogical link between the external environment and the interface.

One of the aims of the investigation reported in Chapter 8 was to assess this extended line of reasoning, and to answer a number of the questions that arose from it. For example, does haptically presented eco-driving support, presented at the site of control, better support eco-driving behaviours than information with equivalent content presented through the auditory or visual channel? Can we extend the argument for combining action and control *surfaces* to combining action and control *sensory modes*? Furthermore, should this successfully support eco-driving, does it do it at the *skill*-based level of cognitive control? In other words, for a manual task (such as depressing an accelerator pedal) is it inherently more supportive of skill-based behaviour to offer information through the manual, that is, haptic, sensory channel?

The first thing to say about the results from Chapter 8 is in relation to the environmental attitudes inventory, the questionnaire chosen to measure the extent to which participants hold more or less pro-environmental attitudes. Not only were there no relationships between scores on this scale and measures of the effects of the information system on driving performance (in any presentation mode or combination), but there was also a total lack of correspondence between environmental attitude scores and driving performance under conditions of 'normal' or 'efficient' driving. Whether this is due to the inadequacy of the tool used to measure environmental awareness, or can be attributed to the disconnect between attitudes and behaviours, is a question we cannot, and do not try to answer in full here; this is a topic to which far more detailed attention is warranted than this concluding chapter can provide. Given the questionnaire's general acceptance in the literature, the well-known disconnect between attitudes and behaviours (e.g. Kollmuss and Agyeman 2002; see also Chapter 3), and Chapter 3's results (showing only weak relationships between environmental attitudes and self-reported driving behaviour), an initial conclusion would be that it is not the tool itself that is to blame. What is important for this research, however, is that we are unable to confirm or discount (using results from this questionnaire) the possibility that vibrotactile information supports behaviour at the skill-based level of cognitive control, thereby bypassing the upper part of the decision ladder (i.e. that which involves goals and motivations).

Some tentative support does, however, come from differences in driving performance at baseline compared to performance under conditions of feedback. After grouping the participants by baseline performance (i.e. more or less economical) it was found that participants of a less economical driving style reacted to visual stimuli, stimuli that have been argued to be more likely to exert control at the rule-based level, to a lesser extent than to haptic information, stimuli that have been argued to be more likely to exert control at the skill-based level. However, auditory information also had a strong effect on all participants; this was also hypothesised to act at the rule-based level. Although one could argue for the success of haptic information over visual from an EID or SRK-informed theoretical standpoint, given the ability of auditory information to foster the same level of compliance one could also argue that these stimuli are simply more salient, thereby encouraging compliance more.

Taken as a whole, it is difficult to confirm or deny, with any authority, the theoretical arguments outlined above. Although it is quite possible to use the EID and SRK theory and principles to justify the use of haptic information in the vehicle, it is difficult to use haptic, in-vehicle information to test EID and SRK theory and principles. Results are in line with theory-based predictions; however, there are other potential explanations. This line of theoretical exploration certainly requires more work, perhaps with a more basic, pure-science approach (as opposed to the applied, simulator-based approach used here).

Despite the lack of a conclusive test of the ability of haptic information, presented at the site of control, to support skill-based acceleration (and deceleration) behaviours, this book nevertheless contributes to the literature surrounding both EID (and the SRK taxonomy), and to the justifications for the use of haptic information in the vehicle. As Chapter 5 showed, the EID literature is heavily biased towards visual interfaces. It is quite possible that to display enough information to support knowledge-based reasoning (through displaying an externalised model of the system; a stipulation of EID) a system *needs* a visual display; indeed, it is difficult to imagine how providing such a display with auditory or haptic information, in a way that is immediately clear, would be possible. Even in Watson et al.'s work, in which EID principles were applied to auditory displays, conventional visual displays were still seen as necessary; the auditory displays were complementary, not a replacement (e.g. Watson and Sanderson 2007).

Moreover, Lee et al. (2004), the only other researchers (to our knowledge) to discuss haptic information in terms of EID or the SRK taxonomy, note that although haptics may well be suited to the support of skill- and rule-based behaviour in the vehicle, to support knowledge-based behaviour with haptic interfaces offers 'the greatest challenge' (Lee et al. 2004, 845). We would argue, however, that this is not problematic; there is no pressing *need* to support knowledge-based reasoning with in-car haptic information, as information in the existing environment (in the outside world and in the in-vehicle displays already present) *already* supports behaviour at this level of cognitive control. Haptic information is useful in supporting behaviour at *lower* levels of cognitive control. As Lee et al. describe, 'haptic interfaces are best suited to support skill and rule-based levels of control, which is precisely what is needed to support drivers' (845, ibid.).

In Chapter 5 the importance of the SRK taxonomy to EID was stressed. Although we do not argue against modifications of the method, or the use of only one part of it (the exploration and modification of methods and approaches is central to the scientific discipline; we simply argue for clear and consistent reporting of its use), we do still maintain that the taxonomy presents a fundamental component, with its omission often presenting a missed opportunity for the guidance of *how* information is to be displayed. Following on from the arguments presented in Chapters 5 and 6 (and the discussion to Chapter 8), we would posit that this is especially true for interfaces that could benefit from the use of information presented in sensory modes other than that of vision. The concepts of combining action and control surfaces (and, relatedly, of allowing the user to act directly on the display), of presenting the boundaries of the system in an immediately perceptible manner (i.e. by taking advantage of the human sensorimotor system), and of providing direct links between the signals provided and the motor responses required, are very much applicable to haptic information presentation, both generally and specifically in the driving domain. Although this book does not present the first exploration of these concepts (see, e.g. Lee et al. 2004), it has expanded upon them considerably.

10.3   METHODOLOGICAL IMPLICATIONS

Before discussing the practical implications of this research, a brief recognition of some of the methodological implications is warranted. The first is with regard to the use of verbal protocol analysis as a tool for exploring differences in cognitive processes between drivers. As was discussed at some length in Chapter 4, no relationships whatsoever could be found between indicators of eco-driving performance and the content of drivers' verbal reports. The possible reasons for this lack of correspondence will not be repeated here (the reader is referred to Chapter 4); however, it does suggest that verbal protocol analysis, in the form argued for by Ericsson and Simon (the originators of the technique; Ericsson and Simon 1980, 1993), is not a method suitable for the analysis of differences between people of similar experience and expertise in a task that can be considered to be highly practised and approaching automaticity.

Second, as will be discussed in more detail in the following section on practical implications, the most commonly reported form of eco-driving tip provided by survey respondents in Chapter 3 was *not* acted on in the simulator study presented in Chapter 8. Although participants *did* reduce their levels of harsh acceleration when asked to drive efficiently (the second most commonly reported tip from the survey results), they did *not* show any additional early actions to upcoming events, a necessary precondition of performing the most commonly reported eco-driving tip in the survey study. This suggests that results from self-reports of driving behaviour should perhaps be taken with caution; knowing about a behaviour having benefits on fuel economy, and stating that it is performed, may not be a reliable indicator of actual in-vehicle behaviour (see also Armitage and Conner (2001) for a discussion on the link between self-reports and observed behaviours).

Third, it is worth discussing the different methods, and categorisation schemes, used in Chapters 3, 4 and 6. In Chapter 3 respondents to an online survey were asked to provide a number of distinct eco-driving tips, in free text. These responses were then coded according to a scheme developed from the eco-driving literature. In Chapter 4 the verbal protocol analysis method was adopted. Here, the aim was to assess whether or not drivers displaying different characteristics (in terms of behaviours indicative of fuel-efficient driving) would also produce verbal reports that reflect differences in underlying cognitive structures or processes. The method (as described by Ericsson and Simon 1993) was followed in a way that matched, as closely as possible, that described by its originators, and was motivated by the goal of learning from more efficient drivers in order to provide support to less efficient drivers. Finally, in Chapter 6, the decision ladder analysis tool was used to model the decision-making processes of expert eco-drivers when performing certain actions. This was rooted in the SRK taxonomy of behaviour, which the decision ladder can be said to represent. The process here was not to categorise that which was reported by the members of the focus group, or the interviewees, but to supplement, alter and ultimately validate the decision ladder models of eco-driving behaviour.

Of course, the selection of a method will depend, to varying degrees, on the study's design, the type of data that has been collected, the goals of the investigation, and even on the expertise and experience held by the principal investigator. For these reasons it would have been inappropriate to try and use the same method or categorisation scheme in the three separate chapters. This would, in fact, have been impossible. Each had different motivations, and each used different types of data.

Categorising the eco-driving tips provided by the respondents to Chapter 3's survey was done in terms of what does and what does not represent a fuel-saving tip that, if followed, could be reasonably assumed to improve economy. This necessarily had to be driven by the eco-driving literature in order to provide justification for a tip's categorisation as 'valid' or 'invalid' (in terms of the tip provided), and to the assignment of a fuel-saving score. This was not the case in Chapter 4, the study using verbal protocol analysis. As described in Chapter 4, the participants were not informed that the data would be investigated from an eco-driving perspective, nor were they asked to focus on any particular aspect of the driving task. Indeed, to do so would have gone against the guidance of Ericsson and Simon, and likely resulted in what they would have argued to be level three verbalisations (i.e. those requiring additional levels of translation or directed attention). The categorisation scheme used, therefore, was developed partly from a scheme used by Banks et al. (2014b), and partly from the verbal protocols themselves. This reflected the largely data-driven approach to the investigation. To have applied the categorisation scheme used to organise responses to the request for eco-driving tips (in Chapter 3) would not only have been inappropriate (given the differences in the types of data), but, given the content of the verbal protocols, would have been impossible. As the results to Chapter 4 showed, there were simply no references whatsoever to eco-driving behaviours.

Finally, in Chapter 6 a third method was adopted, namely the decision ladder model. This chapter specifically aimed at modelling the decision-making processes of expert eco-drivers, in terms of the SRK taxonomy, when performing specific activities. In this sense the data were not categorised at all; the interview process (and the preceding focus group) involved an iterative, participatory effort to populate the diagrams depicting activities in decision-making terms. Once again, the choice of method was dictated by the goals of the investigation.

This being said, it is important to recognise that different methods will provide different perspectives on, or answers to a question. Indeed, different practitioners using the same method, tackling the same issue, may well arrive at different outcomes or solutions. This is particularly true for participatory research, such as the completion of the decision ladders presented in Chapter 6. The method was chosen as it represents a means to describe eco-driving in terms of the SRK taxonomy (in line with the theory behind, and goals of this book); the choice to interview experts was driven by the need to validate the models developed. Should a different researcher have developed those models, and talked with those experts, it is quite likely that the models would have been formulated slightly differently.

This by no means detracts from the value of the analysis. It is akin to the question of design; if one gives two different designers a task, they will almost certainly come out with different solutions. The validity of one solution does not preclude the other from also being valid. They are simply different. Hence although we recognise the somewhat subjective nature of the process, we would argue that it remains useful, and that it retains the potential to offer interesting insights and support design.

Finally, as briefly mentioned in the section on theoretical developments, there is something to be said about the ability of simulator research to answer the fundamental theoretical questions regarding the combination of action and observation sensory modes. Driving simulator research is, of course, more reflective of the real world environment than, for example, a desktop-based study that tests reaction times to arbitrary stimuli presented to various parts of the body. Such a study may, however, be more appropriate for the investigation of the fundamental properties of stimuli presented in different modes. For example, does a spoken stimulus support faster vocal responding than manual responding? Or does the presentation of a vibrotactile stimulus at the site of control (e.g. to the finger for a button click) support faster responding than a vibrotactile stimulus presented to an incongruous site? To answer these questions would likely help to further our basic understanding of human perception and action (questions that, perhaps, are more difficult to answer in the simulator), but would be far less generalisable to the driving domain than experiments conducted in a high-fidelity driving simulator.

10.4   PRACTICAL IMPLICATIONS

10.4.1   INTRODUCTION

One of the reasons for conducting research in the driving simulator (rather than more fundamental, desktop-based research) was that theory was not the only driving force behind the research presented in this book. As well as an exploration of the theoretical implications of the SRK taxonomy, and of the relationship between certain types of stimuli and the behaviours they elicit, the work reported book had a practical motivation; the reduction of energy use (in private transport) via the encouragement of fuel-efficient driving styles. The first step towards achieving such a goal was, of course, to gather information. As Ernst et al. elegantly summarised in relation to software requirements engineering, one must understand 'the terrain before understanding what paths one can take therein' (Ernst et al. 2006, 3); hence the activities described in Chapters 3, 4 and 6.

10.4.2   WHICH BEHAVIOURS TO SUPPORT?

Chapter 3 presented an attempt to first understand the extent to which the general public is aware of eco-driving as a practice, how much people know about it, and their propensity to exhibit eco-driving behaviours. Results of the online survey study suggested that most people are, in fact, aware of eco-driving, and that the great majority of people have a favourable attitude towards it. Although only around half of the respondents were able to provide tips regarding the most influential behaviours (i.e. acceleration and deceleration, and gear choice), the vast majority of respondents were able to provide at least one valid eco-driving tip, with most providing two or more. Results also showed only weak links between an individual's environmental attitudes and their reported propensity to follow their own eco-driving tips; although the links are likely to exist (albeit weakly), it is clear that focusing purely on changing attitudes would not be a sufficient strategy for encouraging behavioural change. The types of tips provided also offered an insight into people's understanding of the practice; the most commonly reported advice was in relation to the avoidance of acceleration and deceleration, closely followed by gentle use of the accelerator and brake pedals, and the efficient use of gears (including minimising time spent at high engine revolutions).

Results from the subsequent chapter (Chapter 4) were far less enlightening. The initial aim was to build an understanding of the potential differences in the cognitive structures held by drivers of differing efficiencies, with the hope of identifying some potential behaviours suitable for support (by in-vehicle information) in drivers of lower fuel efficiency. The total lack of relationships between verbal reports and measurable actions resulted in the conclusion that the verbal protocol analysis methodology was unsuitable, given the aims of the investigation, the groups investigated and the context of use. This was discussed in more detail in the preceding section on methodology; however, here it is sufficient to state that this chapter's results provided no practical input into the eco-driving support system presented in Chapters 8 and 9. For this, a different approach was taken.

One of the possible explanations for the lack of group differences found in Chapter 4 was that the groups were simply not different enough. Therefore, rather than focusing on those slightly more, or slightly less likely to display behaviours characteristic of eco-driving, Chapter 6 looked exclusively at expert eco-drivers. From the five interviews conducted, which themselves built on a review of academic and more publicly available literature, five largely distinct activities, or situations, were identified as having particularly significant effects on fuel economy; headway maintenance, acceleration, deceleration for a full stop, deceleration to a slower speed and deceleration for a road curvature.

Due to limitations of the simulator software (limitations that would have required significant expertise, not available at the time, to overcome), headway maintenance was not included in the in-vehicle information system design considerations. This left three classes of deceleration behaviours (to a slower speed, to a full stop and for a road curvature), and one class of acceleration behaviour (to accelerate to a higher speed, from a lower speed or standstill). The efficient undertaking of all of these activities, confirmed in the interviews and reflective of those found in the literature to be of significance, relies on efficient use of the accelerator pedal; in terms of depression rates for efficient acceleration, and in terms of the timing of its release for efficient deceleration. The four activities could, therefore, be supported by two types of stimuli; one discouraging depression of the pedal beyond 70% (for efficient acceleration), and the other encouraging removal of the foot from the accelerator pedal when approaching an event requiring deceleration (see Chapter 7).

10.4.3 LINKING RESULTS

In terms of the actual behaviour exhibited by participants in the experimental analysis of the eco-driving support system described in Chapter 8, one of the results, taken with results from the aforementioned expert interviews (see also Chapter 6) and online survey (see also Chapter 3), was of particular interest. As described above, the most common eco-driving tip reported by respondents to the survey presented in Chapter 3 was in relation to the avoidance of acceleration and deceleration. Such behaviour is intimately related to the three classes of deceleration behaviour discussed above; the ways in which this avoidance can be achieved include reading the road ahead in order to avoid unnecessarily coming to a stop for road events such as traffic lights (thereby avoiding subsequent unnecessary acceleration), and acting early to changing road conditions in order to avoid usage of the brake pedal (e.g. in anticipation of a road curvature).

One might expect such behaviours, if performed, to be reflected in objective driving measures, such as the distance spent coasting (i.e. travelling without depression of the accelerator pedal). This was, however, the only measure in which *no* significant differences were found between sessions of 'normal' driving and sessions in which participants were asked to drive 'efficiently' (Chapter 8). Although the participants *did* show a reduction in the instances of heavy acceleration when asked to drive efficiently (the 'acceleration' class of behaviour arising from the expert interviews, and the second most commonly reported tip in the responses to the survey), they only showed increased coasting behaviours when provided with additional information encouraging them to do so. This implies that the negative effect of harsh acceleration on fuel economy is known, and performed when driving with fuel-saving goals, whereas the use of coasting as a fuel-saving technique is not.

It is difficult to assert that the general public do not know about early action to slowing events as a method for saving fuel in the vehicle, or that it is a technique reserved for eco-driving experts; the results from the online survey suggest otherwise. It is clear from the results presented in this book, however, that people do not perform such behaviours spontaneously when asked to 'eco-drive'. This, therefore, provided the focus of the second foray into the driving simulator.

10.4.4 FOCUSING ON PRIORITIES

The focus on coasting support was one of two major refinements of the system made following results from the experiment reported in Chapter 8; the other was use of vibrotactile information in isolation, rather than including visual, auditory or any combination thereof. The reason for providing only coasting support stemmed from the fact that this was not spontaneously displayed when asking people to drive efficiently, as previously discussed. In the debriefing interviews briefly outlined in Chapter 8 (conducted after each participant's final driving session had finished) a number of comments were made regarding the confusion of the two stimulus triggers; in certain situations some participants were not always aware whether the stimulus was indicating that they were accelerating excessively, or that they should be beginning a coasting phase. Hence rather than devote in-vehicle stimuli to supporting a behaviour that people already display when driving with the goal of efficiency, the discouragement of excessive acceleration was abandoned as target behaviour for support by the in-vehicle system.

As for the use of vibrotactile information alone (rather than including stimuli presented via any other sensory mode), this was a decision based on two factors; efficacy and acceptance. Regardless of any theoretical arguments (these have been discussed above), the simple finding remains that vibrotactile information fostered greater compliance than did visual information (i.e. participants followed the advice). Moreover, although auditory information also fostered high levels of compliance, this was not at all well received by participants. This is of practical importance insofar as the system can only support reductions in fuel use if it is actually used by drivers.

The system developed for this research project is not one that is envisaged to be activated at all times; on the contrary, it would be something that the driver chooses to engage in situations when efficiency is the primary goal (as opposed to situations involving, e.g. high time pressure; see Chapter 6 for a brief discussion on chosen goals). If the system is greatly disliked (as would likely be the case should auditory stimuli like those assessed here be used), then it is highly unlikely that drivers would use it. Indeed, this line of reasoning also justifies the exclusion of excessive acceleration discouragement. The system would likely be engaged only when drivers have already chosen the goal of driving efficiently; as results from Chapter 8 show, people already display reduced acceleration when performing with this goal. Although a highly sophisticated system informing the driver of the precise foot positioning required at various points along the acceleration profile may indeed help to increase efficiency, the disadvantages (e.g. information overload, confusion) of providing a stimulus for the relatively crude 70% threshold used here (to which participants adhere anyway, without additional information) were judged to outweigh the potential benefits.

10.4.5 Testing Timings

The two reasons for using vibrotactile stimuli in isolation, namely efficacy and acceptance, also provided the motivation for the assessment of different stimulus timings (Chapter 9). In the debrief interviews conducted as part of the study in Chapter 8 a number of participants made comments about the information coming too early (i.e. too far from the event); however, the experts interviewed for the analysis in Chapter 6 all reported acting at the earliest possible opportunity, even if the event were over 500 metres away (reported by two of the five interviewees). These two findings, in combination with results from the literature concerning the timing of coasting advice (Hajek et al. 2011; Staubach et al. 2014b; see Chapter 9), led to the hypothesis that advice presented further from the event to which it refers (i.e. longer lead times) would support greater overall fuel savings, but lower acceptance ratings, whereas information presented closer to the event may be well received by participants, but would result in lower fuel savings.

The results presented in Chapter 9 did not entirely support this hypothesis; although greater benefits (in terms of indicators of efficiency) *were* seen with longer lead times, user acceptance did *not* decline significantly in line with predictions. In general, the medium lead time condition (in which stimuli were presented 8 seconds before the event; the same as in Chapter 8) saw significant improvements over baseline and short lead time; the long lead time condition saw further improvement still. The short lead time stimulus was *not*, however, well received by participants. Its perceived usefulness was significantly lower than medium or long lead-time stimuli, it was significantly *less* satisfying than having no information at all, participants indicated significantly lower behavioural intention to use the system and its timing was considered furthest from ideal.

It is perhaps the case that drivers generally respond more than 4 seconds before an event requiring slowing *anyway*, and offering information at this time point is not only of little use, but may even damage performance. Individuals that would have acted earlier may end up waiting for the advice to be presented, thereby *reducing* coasting distances. Combining this with the low acceptance finding leads us to conclude that stimulus timings of this level are *not* advisable for an in-vehicle information system. Supporting coasting behaviours for fuel efficiency, according to results presented in this book, can have clear benefits to eco-driving performance; however, there is a lower limit.

10.4.6   IN-VEHICLE IMPLEMENTATION

Where exactly the lower limit described above should be is something that will likely vary across vehicles. For example, in electric vehicles where regenerative braking takes effect as soon as the foot is lifted from the accelerator (i.e. without need to depress the brake pedal) the car will decelerate at higher rate than did the vehicle simulated in the experiments presented in this book. This may allow for a lower limit while still being effective (in terms of supporting fuel-efficient driving). Importantly, this system, if implemented in an on-road vehicle, should allow the driver to choose the setting. Some drivers preferred the longer lead-time stimulus, others preferred the medium lead-time stimulus, and others still would have liked even greater lead times. It may be suitable, therefore, to design a system that has, for example, two or three settings, ranging from a 'mild' setting (one that presents information only slightly further from the event than the point at which most drivers would act anyway) to a setting that encourages much greater coasting distances than would normally be seen.

10.5   FUTURE WORK

One obvious limitation to the practical aspect of this research is that the system developed has only been tested in a driving simulator; its effects have not been investigated in an on-road environment. The Southampton University Driving Simulator is fixed-based, and although it simulates vehicle noise, it does not have the capacity to simulate the vibrations present when driving on real roads. How these road vibrations might interact or interfere with, or even mask the additional vibrations investigated in this book remains unanswered. We would posit that the vibrations presented in the accelerator pedal would likely be of a different frequency and intensity to those present in the vehicle. Additionally, we would argue that the two would likely be distinguishable due to the fact that the foot is not the only site through which road vibrations are felt; rather they are also felt through the seat and the steering wheel, in a comparable (if not equal) intensity and frequency. The foot would be the only site to receive vibration of a different frequency and intensity to that received by other body areas. This would, however, require further investigation.

There also remains the question of whether or not the system described in this book would also be suitable for use in vehicles with regenerative braking, and to what extent it could help increase efficiency in these vehicles. The initial guiding aim of this book was to inform interface design in low-carbon vehicles. Although this broadened to include the support of fuel-efficient behaviour in any vehicle, the question of how best to support efficient driving behaviours in vehicles that have different acceleration and deceleration characteristics remains. We would suggest that to present to the driver a vibrotactile stimulus, through the accelerator pedal, informing them of the optimum time to remove the foot from said pedal when approaching deceleration events would still likely result in reductions in fuel consumption. It is highly likely, however, that the timing of the advice would need to be different. As has been discussed, the most efficient time ahead of an event to start the coasting phase will depend (among other things) on the characteristics of the vehicle, specifically its engine braking characteristics. Regenerative braking increases the rate at which a vehicle decelerates once the foot has been lifted from the accelerator pedal, hence the number of seconds ahead of the event at which the stimulus is provided (compared to the timings investigated in Chapters 8 and 9) would almost certainly need to be reduced.

Relatedly, it may be the case that in hybrid vehicles a more holistic form of acceleration and deceleration support is more appropriate. Rather than simply encouraging enhanced coasting behaviours, the driver could be guided in all their accelerator pedal use. In Chapter 2 research from Franke et al. (2016) was introduced, and the following quote presented:

> '... drivers suggested that certain critical system states should be more clearly displayed (e.g. the point of maximum efficiency of the combustion engine, the neutral point at which there is zero energy flow in the system, the point at which regenerative braking is optimal, or a point just before that at which the combustion engine turns on), and that targeting these points should be facilitated.' (Franke et al. 2016, 39)

This comes from an investigation of hybrid vehicle users specifically, and describes the kind of support that experienced users of the technology requested when asked about the types of devices that they think could support them in maximising their fuel efficiency (Franke et al. 2016). Indeed, 18% of the 39 drivers interviewed went so far as to suggest that detailed haptic or vibrotactile feedback, presented through the accelerator pedal, might be a suitable means to present the various points described in the above quote. This is clearly merits further research; not only has this book shown such information to offer benefits, but users of the technology have actively requested such support.

To continue with a discussion of the support of all forms of accelerator pedal usage, another avenue for future work stems from the possibility that the line of thinking on the abandonment of the discouragement of excessive acceleration is faulty. We have argued that the avoidance of excessive acceleration is not as worthy of support by the in-vehicle system as is enhanced coasting; however, it is quite possible that rather than spontaneously exhibiting reductions in acceleration (as participants in Chapter 8 did) upon activation of a fuel-saving goal (a necessary precondition for turning on an in-vehicle eco-driving system), the participants rely wholly on the system to guide their acceleration and deceleration behaviours. This might see people accelerating excessively, as there would be no cue to tell them otherwise. Although the system would support only one aspect of eco-driving (i.e. coasting), the participant might expect the system to support eco-driving as a whole, that is, giving information on all accelerator pedal movements. A potential study could, therefore, compare the effects of a tool that provides only coasting support, with one that offers both coasting and reduced acceleration support.

The vibrotactile stimulus used in this research here was binary, that is, either on or off. It may be interesting to assess how a graded or phased stimulus affects participant performance and acceptance. For example, rather than simply turning on at a particular point on the road, vibration intensity could gradually increase as the vehicle nears an event. This may also be used in conjunction with the acceleration support described above; if the stimulus onset is more subtle, it may help people to experiment with the system, 'feeling' the optimum acceleration profile in a more subtle way, thereby producing a more idealised acceleration and deceleration profile over the whole journey. This may reduce the confusion reported by some participants in Chapter 8, and support both aspects of accelerator pedal usage.

Another possible means for reducing any potential confusion, while still retaining support for both behaviours, is to provide the two types of advice (i.e. reduced acceleration, increased coasting) in different sensory modes. It appears that the vibrotactile stimulus works well for coasting support; however, it may also be possible to support smoother acceleration with an auditory stimulus. This idea is of particular interest in relation to purely electric vehicles, in which little or no engine noise is present. Currently, in a car powered by an internal combustion engine, it is possible to hear when the engine is at high revolutions, one symptom of excessively harsh acceleration. Such a stimulus, already familiar to the driver as an indicator of inefficient use of the engine, could be replicated (and, indeed, modified) and added to the in-vehicle environment, perhaps in a way that increases gradually as the driver depresses the accelerator pedal. Whether this noise would be better as a replication of engine noise, or some other, completely distinct sound, is also of interest.

To return to the investigation of a system focusing only on haptic stimuli, there are a number of possible extensions and comparisons that could be made. For example, in Chapter 6 one of the identified eco-driving activities was headway maintenance. Due to simulator limitations the potential for supporting this behaviour was not investigated. If the same approach as that described above were adopted (i.e. gradually increasing stimulus intensities as the event nears) it could be possible to support longer headway distances as well as enhanced coasting (and, quite possibly, reduction in harsh acceleration as well). If all alerts were included, such a system could guide the user's accelerator pedal usage almost entirely. All that would be required from the user is to 'feel' for the point at which pedal depression is optimal (a decision made via the integration of various sources of information, e.g. vehicle engine parameters, digital and topographical mapping data, radar – see Continental's (2015) eHorizon project, introduced in Chapter 1). This kind of system could also incorporate the optimum accelerator profile for hill driving, the negotiation of which can have a significant effect on fuel economy (e.g. Schwarzkopf and Leipnik 1977).

Returning to the specific question of how to best support enhanced coasting behaviours for maximum efficiency, the issue of stimulus timing would be worthy of further attention. In general, the participants in the study described in Chapter 9 gave favourable ratings to both the 8 and 12 second lead time conditions. This raises the question; how far in advance of an event are people willing to accept coasting advice? Depending on the vehicle, it may be beneficial (in terms of fuel economy) to remove the foot from the pedal as far as (or even exceeding) 500–600 metres from an event. At what point does diminishing user acceptance outweigh the potential fuel-efficiency benefits? Relatedly, it would be interesting to assess a system that allows the user to select the stimulus lead time themselves; indeed, it is our opinion that if coasting support is to be added to the vehicle, it should be in a form that allows the user to set their own parameters (down to a lower limit; see Chapter 9 and the discussion on stimulus timing earlier in this chapter).

Although this research focused on vibrotactile feedback in the vehicle, the great majority of research into supporting coasting, or indeed haptic feedback presented through the accelerator pedal for any purpose, uses counter force or added stiffness (e.g. Adell et al. 2008; Hajek et al. 2011; Mulder et al. 2010, 2011; Staubach et al. 2014a, 2014b). To our knowledge, there exists no published research on the comparison of the effects, on both driving performance and user acceptance, of these two types of stimuli. Such a study would be highly worthwhile.

One final question (at least in terms of the practical considerations of fuel use, participant acceptance and system efficacy) that is of potential interest is the effect on the traffic system as a whole that an in-vehicle system such as this would have; should some drivers maximise coasting phases and others not, what would this do to the system? The research presented in this book focuses on the individual driver; however, it is important to recognise that each one is part of a larger, more complex, socio-technical system. To answer such a question would, we suspect, require a wholly distinct approach to that which has been used in this research project, and is a question that we leave entirely open.

The above suggestions and questions all pertain to the practical aspect of this book. There are, of course, theoretical questions that have as yet been left unanswered. The possibility of following a more basic, pure scientific approach to the study of the effects of congruous and incongruous stimulus presentation and control action sites and sensory modes was discussed above. For example, to assess whether vibration presented at the finger encourages quicker push-button response timings than vibrations presented to other bodily sites. Or, to measure verbal reaction times to a verbal stimulus, compared to reactions made manually. It may also be possible, however, to retain the benefits of the applied, simulator-based approach while also assessing the theoretical arguments surrounding the combination of action and control surfaces. For example, to assess whether these arguments also apply to vibrotactile stimuli it would be interesting to provide a stimulus presented through the accelerator pedal (as has been done in this book) and compare its effects to a vibro-tactile stimulus, requiring the same action (i.e. removal of the foot from the pedal), presented through the steering wheel or seat. Such a study would be relatively simple to design, but would require sensitive measures of reaction time, a measure that has not been used in the research presented in this book. Indeed, such a measure was sought; however, technical limitations prevented its use. It would likely have shed further light on the findings presented in Chapter 8, and as such its omission is accepted as a limitation of this research. Indeed, the theoretical questions that this book has made an early attempt at addressing are left very much open. It would be our hope that future research explores these issues more deeply, the results of which would likely forward our general understanding of cognition and action, an under-standing that would perhaps help us to design better interfaces overall, in the vehicle and beyond.

10.6 CONCLUSIONS

It is impossible to single out any one policy, educational strategy or technological intervention that will make people care enough about the environment in which they live to change their long-held habits and behaviours. Unfortunately, this is exactly what we need to do as a society if we are avoid the most damaging effects of our culture of consumption and disposal. Without action on the part of the individual consumer, there can be little hope of business or industry changing its practices. Regardless of whether one 'believes' in human-induced climate change or not, that we are using more resources than the planet has the capacity to replenish is an ines-capable reality. To claim that eco-driving is the solution to all our worldly problems would be fantasy; its effect on global emissions volumes, though significant, is rela-tively small when compared with the amount saved by, for example, switching from coal to wind as a source of power. It is, however, a behavioural change that costs little, incurs only slight increases to journey times (according to results in this book at least), and can be done by a vast number of people across the globe.

We would, of course, very much like to see a system similar to the one presented in this book implemented in commercially available, on-road vehicles. Indeed, the increasing trend in research regarding haptic feedback, particularly in that which relates to fuel efficiency, and the burgeoning interest in concepts such as Continental's eHorizon, in which the vehicle 'knows' what will be coming up in the road ahead, are both promising in this respect. Although the problem of providing motivation to use such a system remains, at least we will be able to optimally support those that *do* wish to drive efficiently.

References

Abbink, D. A. (2006). *Neuromuscular analysis of haptic gas pedal feedback during car following*. Technische Universiteit Delft. Delft, The Netherlands.

Abbink, D. A., Boer, E. R., & Mulder, M. (2008). Motivation for continuous haptic gas pedal feedback to support car following. In *2008 IEEE Intelligent Vehicles Symposium* (pp. 283–290).

Åberg, L., Larsen, L., & Beilinsson, L. (1997). Observed vehicle speed and drivers' perceived speed of others. *Applied Psychology, 46*(3), 287–302.

Abrahamse, W., Steg, L., Vlek, C., & Rothengatter, T. (2005). A review of intervention studies aimed at household energy conservation. *Journal of Environmental Psychology, 25*(3), 273–291.

Adell, E., & Várhelyi, A. (2008). Driver comprehension and acceptance of the active accelerator pedal after long-term use. *Transportation Research Part F: Traffic Psychology and Behaviour, 11*(1), 37–51.

Adell, E., Várhelyi, A., & Hjälmdahl, M. (2008). Auditory and haptic systems for in-car speed management – A comparative real life study. *Transportation Research Part F: Traffic Psychology and Behaviour, 11*(6), 445–458.

Ahlstrom, U. (2005). Work domain analysis for air traffic controller weather displays. *Journal of Safety Research, 36*, 159–169.

Ajzen, I. (1991). The theory of planned behavior. *Organizational Behavior and Human Decision Processes, 50*, 179–211.

Alicke, M. D. (1985). Global self-evaluation as determined by the desirability and controllability of trait adjectives. *Journal of Personality and Social Psychology, 49*(6), 1621–1630.

Alicke, M. D., Breitenbecher, D. L., Yurak, T. J., & Vredenburg, D. S. (1995). Personal contact, individuation, and the better-than-average effect. *Journal of Personality and Social Psychology, 68*(5), 804–825.

Alsabaan, M., Naik, K., & Khalifa, T. (2013). Optimization of fuel cost and emissions using V2V communications. *IEEE Transactions on Intelligent Transportation Systems, 14*(3), 1449–1461.

Altman, D. G. (1991). *Practical statistics for medical research*. London: Chapman and Hall.

Amelink, M. H. J., Mulder, M., van Paassen, M. M., & Flach, J. (2005). Theoretical foundations for a total energy-based perspective flight-path display. *The International Journal of Aviation Psychology, 15*, 205–231.

Amelink, M. H. J., van Paassen, M. M., Mulder, M., & Flach, J. M. (2003). Applying the abstraction hierarchy to the aircraft manual control task. In *Proceedings of the 12th International Symposium on Aviation Psychology* (pp. 1–6). April 14–17. Dayton, OH.

Ammons, R. B. (1956). Effects of knowledge of performance: A survey and tentative theoretical formulation. *Journal of General Psychology, 50*, 279–299.

Anderson, J. R. (1976). *Language, memory, and thought*. Hillsdale, NJ: Erlbaum.

Anderson, J. R. (1983). *The architecture of cognition*. Cambridge, MA: Harvard University Press.

Anderson, J. R. (1993). *Rules of the mind*. Hillsdale, NJ: Erlbaum.

Arcury, T. (1990). Environmental attitude and environmental knowledge. *Human Organization, 49*(4), 300–304.

Armitage, C. J., & Conner, M. (2001). Efficacy of the theory of planned behaviour: A meta-analytic review. *The British Journal of Social Psychology, 40*(4), 471–499.

Arroyo, E., Sullivan, C., & Selker, T. (2006). CarCoach: A polite and effective driving coach. In *Proceedings of the CHI: Conference on Human Factors in Computing Systems*, Montreal, Canada, April 22–27, (pp. 357–362).

Atkinson, R. C., & Shiffrin, R. M. (1968). Human memory: A proposed system and its control processes. *The Psychology of Learning and Motivation, 2*, 89–195.

Automotive Design. (2010). 60 second interview with J. Mays. *Automotive Design*, May 2010, p. 35.

Azzi, S., Reymond, G., Mérienne, F., & Kemeny, A. (2011). Eco-driving performance assessment with in-car visual and haptic feedback assistance. *Journal of Computing and Information Science in Engineering, 11*(4), 041005.1–041005.5.

Baddeley, A. D., & Hitch, G. J. (1974). Working memory. *The Psychology of Learning and Motivation, 8*, 47–89.

Bamberg, S., & Möser, G. (2007). Twenty years after Hines, Hungerford, and Tomera: A new meta-analysis of psycho-social determinants of pro-environmental behaviour. *Journal of Environmental Psychology, 27*(1), 14–25.

Bandura, A. (1986). *Social foundations of thought and action: A social cognitive theory.* Englewood Cliffs, NJ: Prentice-Hall.

Banks, V. A., Stanton, N. A., & Harvey, C. (2014a). Sub-systems on the road to vehicle automation: Hands and feet free but not 'mind' free driving. *Safety Science, 62*, 505–514.

Banks, V. A., Stanton, N. A., & Harvey, C. (2014b). What the drivers do and do not tell you: Using verbal protocol analysis to investigate driver behaviour in emergency situations. *Ergonomics, 57*(3), 332–342.

Barkenbus, J. N. (2010). Eco-driving: An overlooked climate change initiative. *Energy Policy, 38*(2), 762–769.

Barth, M., & Boriboonsomsin, K. (2008). Real-world CO_2 impacts of traffic congestion. *Transportation Research Record: Journal of the Transportation Research Board, 2058*, 163–171.

Bartlett, F. C. (1932). *Remembering: A study of experimental and social psychology.* Cambridge: Cambridge University Press.

Beckx, C., Int Panis, L., De Vlieger, I., & Wets, G. (2007). Influence of gear changing behaviour on fuel-use and vehicular exhaust emissions. In *Highway and urban environment* (pp. 45–51). Eds., Morrison, G.M., Rauch, Sébastien. Netherlands: Springer.

Beevis, D., Vicente, K. J., & Dinadis, N. (1998). An exploratory application of ecological interface design to aircraft systems. In *RTO HFM Symposium on Collaborative Performance in Complex Operational Systems* (pp. 2-1–2-9). Edinburgh, UK.

Bennett, K. B., & Flach, J. M. (2011). *Display and interface design. Subtle science, exact art.* Boca Raton, FL: CRC Press.

Bennett, K. B., Posey, S. M., & Shattuck, L. G. (2008). Ecological interface design for military command and control. *Journal of Cognitive Engineering and Decision Making, 2*(4), 349–385.

Berkhout, P. H. G., Muskens, J. C., & Velthuijsen, J. W. (2000). Defining the rebound effect. *Energy Policy, 28*, 425–432.

Berry, M. (2013). Hard evidence: How biased is the BBC? *The Conversation*. Retrieved January 15, 2016, from https://theconversation.com/hard-evidence-how-biased-is-the-bbc-17028

Bertin, J. (1983). *Semiology of graphics: Diagrams, networks, maps.* Madison, WI: The University of Wisconsin Press.

Best, H., & Lanzendorf, M. (2005). Division of labour and gender differences in metropolitan car use. An empirical study in Cologne, Germany. *Journal of Transport Geography, 13*(2), 109–121.

Bilodeau, E. A., & Bilodeau, I. M. (1961). Motor-skills learning. *Annual Review of Psychology, 12*, 243–280.

Bingham, C., Walsh, C., & Carroll, S. (2012). Impact of driving characteristics on electric vehicle energy consumption and range. *IET Intelligent Transport Systems, 6*(1), 29–35.

Birrell, S. A., Fowkes, M., & Jennings, P. A. (2014). Effect of using an in-vehicle smart driving aid on real-world driver performance. *IEEE Transactions on Intelligent Transportation Systems, 15*(4), 1801–1810.

Birrell, S. A., Taylor, J., McGordon, A., Son, J., & Jennings, P. (2014). Analysis of three independent real-world driving studies: A data driven and expert analysis approach to determining parameters affecting fuel economy. *Transportation Research Part D: Transport and Environment, 33*, 74–86.

Birrell, S. A., & Young, M. S. (2011). The impact of smart driving aids on driving performance and driver distraction. *Transportation Research Part F: Traffic Psychology and Behaviour, 14*(6), 484–493.

Birrell, S. A., Young, M. S., & Jenkins, D. P. (2008). Improving driver behaviour by design: A cognitive work analysis methodology. In *2nd International Conference on Applied Human Factors and Ergonomics.* July 14–17, Las Vegas, NV.

Birrell, S. A., Young, M. S., Jenkins, D. P., & Stanton, N. A. (2012). Cognitive work analysis for safe and efficient driving. *Theoretical Issues in Ergonomics Science, 13*, 430–449.

Birrell, S. A., Young, M. S., & Weldon, A. M. (2013). Vibrotactile pedals: Provision of haptic feedback to support economical driving. *Ergonomics, 56*(2), 282–292.

Bisantz, A. M., & Vicente, K. J. (1994). Making the abstraction hierarchy concrete. *International Journal of Human-Computer Studies, 40*, 83–117.

Blaauw, G. J. (1982). Driving experience and task demand in simulator and instrumented car: A validation study. *Human Factors, 24*, 473–486.

Block, F. E., Nuutinen, L., & Ballast, B. (1999). Optimization of alarms: A study on alarm limits, alarm sounds, and false alarms, intended to reduce annoyance. *Journal of Clinical Monitoring and Computing, 15*(2), 75–83.

Borst, C., Sjer, F. A., Mulder, M., van Paassen, M. M., & Mulder, J. A. (2008). Ecological approach to support pilot terrain awareness after total engine failure. *Journal of Aircraft, 45*(1), 159–171.

Borst, C., Suijkerbuijk, H. C. H., Mulder, M., & van Paassen, M. M. (2006). Ecological interface design for terrain awareness. *The International Journal of Aviation Psychology, 16*, 375–400.

Brunswik, E. (1956). *Perception and the representative design of psychological experiments* (2nd ed.). Berkeley, CA: University of California Press.

Burgess, M., Harris, P. M., Mansbridge, S., Lewis, E., & Walsh, C. (2011). *Initial findings from the ultra low carbon vehicle demonstrator programme – How quickly did users adapt?* Retrieved April 5, 2012, from http://www.innovateuk.org/_assets/pdf/press-releases/ulcv_reportaug11.pdf

Burns, C. M. (1999). Scanning patterns with ecological displays when abstraction levels are separated. In *Proceedings of the Human Factors and Ergonomics Society 43rd Annual Meeting*, Westin Galleria, Houston, Texas, September 27–October 1 (pp. 163–167).

Burns, C. M. (2000a). Errors in searching for abstraction hierarchy information. In *Proceedings of the Human Factors and Ergonomics Society Annual Meeting*, in conjuction with the International Ergonomics Association XIVth Triennial Congress, San Diego Marriott Hotel, July 29–August 4 (Vol. 44, pp. 270–273).

Burns, C. M. (2000b). Navigation strategies with ecological displays. *International Journal of Human-Computer Studies, 52*(1), 111–129.

Burns, C. M. (2000c). Putting it all together: Improving display integration in ecological displays. *Human Factors: The Journal of the Human Factors and Ergonomics Society, 42*(2), 226–241.

Burns, C. M., Garrison, L., & Dinadis, N. (2003). From analysis to design: WDA for the petrochemical industry. *Proceedings of the Human Factors and Ergonomics Society Annual Meeting, 47*(3), 258–262.

Burns, C. M., & Hajdukiewicz, J. (2004). *Ecological interface design.* Boca Raton, FL: CRC Press.

Burns, C. M., Kuo, J., & Ng, S. (2003). Ecological interface design: A new approach for visualizing network management. *Computer Networks, 43*(3), 369–388.

Burns, C. M., Momtahan, K., & Enomoto, Y. (2006). Supporting the strategies of cardiac nurse coordinators using cognitive work analysis. *Proceedings of the Human Factors and Ergonomics Society Annual Meeting, 50*(3), 442–446.

Burns, C. M., Skraaning, G., Jamieson, G. A., Lau, N., Kwok, J., Welch, R., & Andresen, G. (2008). Evaluation of ecological interface design for nuclear process control: Situation awareness effects. *Human Factors: The Journal of the Human Factors and Ergonomics Society, 50*(4), 663–679.

Campbell-Hall, V., & Dalziel, D. (2011). *Eco-driving: Factors that determine take-up of post-test training research.* TNS-BMRB report v5210046. https://www.gov.uk/government/uploads/system/uploads/attachment_data/file/142536/Eco_safe_driving.pdf (Accessed December 20, 2014).

Canfield, K., & Petrucci, K. (1993). Interface design for clinical information systems: An ecological interface design approach. In *Human Computer Interaction: Vienna Conference, VCHCI'93, Fin de Siècle Vienna, Austria, September 20–22, 1993 Proceedings* (pp. 391–402). Berlin: Springer.

Carrasco, C., Jamieson, G. A., & St-Cyr, O. (2014). Revisiting three ecological interface design experiments to investigate performance and control stability effects under normal conditions. In *2012 IEEE Conference on Systems, Man, and Cybernetics*, October 5–8, Paradise Point Resort & Spa San Diego, CA (pp. 329–334).

Chery, S., Vicente, K. J., & Farrell, P. S. E. (1999). Perceptual control theory and ecological interface design: Lessons learned from the control display unit. In *Proceedings of the Human Factors and Ergonomics Society 43rd Annual Meeting*, Westin Galleria, Houston, Texas, September 27–October 1 (Vol. 43, pp. 389–393).

Christoffersen, K., Hunter, C. N., & Vicente, K. J. (1994). Ecological interface design and operator competencies: A (very) long-term study. *Proceedings of IEEE International Conference on Systems, Man and Cybernetics*, San Antonio, TX, October 2–5 (Vol. 2, pp. 1398–1403).

Christoffersen, K., Hunter, C. N., & Vicente, K. J. (1998). A longitudinal study of the effects of ecological interface design on deep knowledge. *International Journal of Human-Computer Studies, 48*, 729–762.

Cleveland, W. P., Fleming, E. S., & Lee, G. (2011). TCAS traffic display redesign. In *2011 IEEE Systems and Information Engineering Design Symposium*, University of Virginia Charlottesville, Virginia, April 29 (pp. 209–214).

Cocron, P., Bühler, F., Neumann, I., Franke, T., Krems, J. F., Schwalm, M., & Keinath, A. (2011). Methods of evaluating electric vehicles from a user's perspective – The MINI E field trial in Berlin. *IET Intelligent Transport Systems, 5*(2), 127–133.

Cohen, J. (1988). *Statistical power analysis for the behavioral sciences* (2nd ed.). Hillsdale, NJ: Lawrence Erlbaum Associates.

Commission for Integrated Transport. (2007). *Transport and Climate Change.* The Stationary Office, London.

Conner, M., Smith, N., & McMillan, B. (2003). Examining normative pressure in the theory of planned behaviour: Impact of gender and passengers on intentions to break the speed limit. *Current Psychology, 22*(3), 252–263.

Continental. (2015). *eHorizon.* Retrieved November 25, 2015, from http://www.continental-automotive.com/www/automotive_de_en/themes/passenger_cars/interior/connectivity/pi_ehorizon_en.html

Cook, J., Nuccitelli, D., Green, S. A., Richardson, M., Winkler, B., Painting, R., Way, R., Jacobs, P., & Skuce, A. (2013). Quantifying the consensus on anthropogenic global warming in the scientific literature. *Environmental Research Letters, 8*(2), 024024.

Corbett, A. T., & Anderson, J. R. (2001). *Locus of feedback control in computer-based tutoring: Impact on learning rate, achievement and attitudes.* In J. Jacko, A. Sears, M. Beaudouin-Lafon, & R. Jacob (Eds.), *Proceedings of ACM CHI 2001 Conference on Human Factors in Computing Systems* , March 31–April 05 (pp. 245–252). New York: ACM Press.

Corral-Verdudo, V. (1997). Dual 'realities' of conservation behavior: Self-reports vs observations of re-use and recycling behavior. *Journal of Environmental Psychology, 17*(2), 135–145.

Cronbach, L. J. (1951). Coefficient alpha and the internal structure of tests. *Psychometrika, 16*, 297–334.

Cronbach, L. J. (1975). Five decades of public controversy over mental testing. *American Psychologist, 1*, 1–14.

Cummings, M. L., & Guerlain, S. (2003). The tactical tomahawk conundrum: Designing decision support systems for revolutionary domains. In *IEEE Systems, Man, and Cybernetics Society Conference*, October 8 (pp. 1583–1588). Washington, DC.

Cummings, M. L., Guerlain, S., & Bass, E. J. (2004). Informing design of a command and control decision support interface through an abstraction hierarchy. In *Proceedings of the Human Factors and Ergonomics Society 48th Annual Meeting*, Sheraton New Orleans Hotel, New Orleans, Louisiana, September 20–24 (Vol. 48, pp. 489–493).

Dainoff, M. J., Dainoff, C. A., & McFeeters, L. (2004). On the application of cognitive work analysis to the development of a commercial investment software tool. In *Proceedings of the Human Factors and Ergonomics Society 48th Annual Meeting*, Sheraton New Orleans Hotel, New Orleans, Louisiana, September 20–24 (Vol. 48, pp. 595–599).

Damiani, S., Deregibus, E., & Andreone, L. (2009). Driver-vehicle interfaces and interaction: Where are they going? *European Transport Research Review, 1*(2), 87–96.

Davies, T. C., Burns, C. M., & Pinder, S. D. (2006). Using ecological interface design to develop an auditory interface for visually impaired travellers. In *OZCHI 2006*, Sydney, Australia, November 20–24 (pp. 309–312). ACM Press.

Davies, T. C., Burns, C. M., & Pinder, S. D. (2007). Testing a novel auditory interface display to enable visually impaired travelers to use sonar mobility devices effectively. In *Proceedings of the Human Factors and Ergonomics Society 51st Annual Meeting*, Baltimore Marriott Waterfront Hotel, Baltimore, Maryland, October 1–5 (pp. 278–282).

De Rosario, H., Louredo, M., Díaz, I., Soler, A., Gil, J. J., Solaz, J. S., & Jornet, J. (2010). Efficacy and feeling of a vibrotactile frontal collision warning implemented in a haptic pedal. *Transportation Research Part F: Traffic Psychology and Behaviour, 13*(2), 80–91.

De Vlieger, I. (1997). On board emission and fuel consumption measurement campaign on petrol-driven passenger cars. *Atmospheric Environment, 31*(22), 3753–3761.

Delhomme, P., Chappé, J., Grenier, K., Pinto, M., & Martha, C. (2010). Reducing air-pollution: A new argument for getting drivers to abide by the speed limit? *Accident Analysis and Prevention, 42*(1), 327–338.

Delhomme, P., Cristea, M., & Paran, F. (2013). Self-reported frequency and perceived difficulty of adopting eco-friendly driving behavior according to gender, age, and environmental concern. *Transportation Research Part D: Transport and Environment, 20*, 55–58.

Department for Transport. (2014). *National travel survey: 2013*. Retrieved October 5, 2014, from https://www.gov.uk/government/statistics/national-travel-survey-2013

Department of Energy and Climate Change. (2012a). *Digest of United Kingdom energy statistics*. London: The Stationary Office.

Department of Energy and Climate Change. (2012b). *Statistical release: 2011 UK green-house gas emissions, provisional figures and 2010 UK greenhouse gas emissions, final figures by fuel type and end-user.* The Stationary Office, London.

Devon County Council. (2013). *Ecodriving film.* Retrieved April 12, 2013, from http://www.devon.gov.uk/index/video/videotransport/ecodrivingvid.htm

Diamantopoulos, A., Schlegelmilch, B. B., Sinkovics, R. R., & Bohlen, G. M. (2003). Can socio-demographics still play a role in profiling green consumers? A review of the evidence and an empirical investigation. *Journal of Business Research, 56*(6), 465–480.

Dinadis, N., & Vicente, K. J. (1995). Does ecological interface design scale up to industrial plants? In *1995 IEEE International Conference on Systems, Man and Cybernetics. Intelligent Systems for the 21st Century,* October 22–25 Vancouver, BC, Canada (Vol. 4, pp. 3133–3138).

Dinadis, N., & Vicente, K. J. (1996). Ecological interface design for a power plant feedwater subsystem. *IEEE Transactions on Nuclear Science, 43*(1), 266–277.

Dinadis, N., & Vicente, K. J. (1999). Designing functional visualizations for aircraft systems status displays. *The International Journal of Aviation Psychology, 9,* 241–269.

Donmez, B., Boyle, L. N., & Lee, J. D. (2007). Safety implications of providing real-time feedback to distracted drivers. *Accident, Analysis and Prevention, 39*(3), 581–590.

Downar, J., Crawley, A. P., Mikulis, D. J., & Davis, K. D. (2002). A cortical network sensitive to stimulus salience in a neutral behavioral context across multiple sensory modalities. *Journal of Neurophysiology, 87*(1), 615–620.

Drivalou, S. (2005). Supporting critical operational conditions in an electricity distribution control room through ecological interfaces. In *EACE '05 Proceedings of the 2005 Annual Conference on European Association of Cognitive Ergonomics,* Chania, Greece, September 29–October 01 (pp. 263–270).

Drivalou, S. (2008). Building-up cognitive artifacts for a complex socio-technical system. In *Proceedings of Applied Human Factors and Ergonomics International Conference 2008.* July 14–17, Las Vegas, Nevada.

Drivalou, S., & Marmaras, N. (2003). Tracing interface design solutions for an electricity distribution network control system using the abstraction hierarchy. In *Proceedings of the International Ergonomics Association XVth Triennial Congress.* August 24–29, Seoul, Korea.

Drivalou, S., & Marmaras, N. (2006). Retrofitting artefacts. In *Proceedings of the IEA 2006.* Maastricht, The Netherlands.

Drivalou, S., & Marmaras, N. (2009). Supporting skill-, rule-, and knowledge-based behaviour through an ecological interface: An industry-scale application. *International Journal of Industrial Ergonomics, 39*(6), 947–965.

Driver and Vehicle Standards Agency. (2015). *The official DVSA guide to driving.* London: The Stationary Office.

Duez, P. P. (2003). *Testing the generalizability of ecological interface design to computer network monitoring.* University of Toronto. Toronto, Canada.

Duez, P. P., & Vicente, K. J. (2003). Ecological interface design for network management. In *Proceedings of the Human Factors and Ergonomics Society 47th Annual Meeting,* Adam's Mark Hotel, Denver, Colorado, October 13–17 (Vol. 47, pp. 572–575).

Duez, P. P., & Vicente, K. J. (2005). Ecological interface design and computer network management: The effects of network size and fault frequency. *International Journal of Human-Computer Studies, 63*(6), 565–586.

Ecowill. (2015). *The golden rules of ecodriving.* Retrieved December 22, 2015, from http://www.ecodrive.org/en/what_is_ecodriving-/the_golden_rules_of_ecodriving

Edmunds.com. (2009). *Hypermiling: The quest for ultimate fuel economy.* https://www.edmunds.com/fuel-economy/hypermiling-quest-for-ultimate-fuel-economy.html (Accessed February 15, 2016).

Effken, J. A. (2006). Improving clinical decision making through Ecological Interfaces. *Ecological Psychology, 18*, 283–318.

Effken, J. A., Johnson, K., Loeb, R., Johnson, S., & Reyna, V. (2002). Evaluating ecologically designed clinical information displays for trauma management. *Health IT Advisory Report, 3*, 3–7.

Effken, J. A., Loeb, R. G., Johnson, K., Johnson, S., & Reyna, V. (2002). Preliminary evaluation of an oxygenation management display. In N. Callaos, L. Hernandez-Encinas, & F. Yetim (Eds.), *Proceedings of the 6th World Multiconference on Systemics, Cybernetics, and Informatics: Vol. 1 Information Systems Development I* (pp. 279–284). Orlando, FL: International Institute of Informatics and Systemics.

Effken, J. A., Loeb, R. G., Kang, Y., & Lin, Z. (2008). Clinical information displays to improve ICU outcomes. *International Journal of Medical Informatics, 77*, 765–777.

Ellerbroek, J., Brantegem, K. C. R., van Paassen, M. M., de Gelder, N., & Mulder, M. (2013). Experimental evaluation of a coplanar airborne separation display. *IEEE Transactions on Human-Machine Systems, 43*(3), 290–301.

Ellerbroek, J., Brantegem, K. C. R., van Paassen, M. M., & Mulder, M. (2013). Design of a coplanar airborne separation display. *IEEE Transactions on Human-Machine Systems, 43*(3), 277–289.

Ellerbroek, J., Visser, M., Van Dam, S. B. J., Mulder, M., & van Paassen, M. M. (2011). Design of an airborne three-dimensional separation assistance display. *IEEE Transactions on Systems, Man, and Cybernetics – Part A: Systems and Humans, 41*, 863–875.

Energy Saving Trust. (2015). *Drive smarter.* Retrieved December 22, 2015, from http://www.energysavingtrust.org.uk/domestic/drive-smarter

Engels, A., Hüther, O., Schäfer, M., & Held, H. (2013). Public climate-change skepticism, energy preferences and political participation. *Global Environmental Change, 23*(5), 1018–1027.

Enomoto, Y., Burns, C. M., Momtahan, K., & Caves, W. (2006). Effects of visualization tools on cardiac telephone consultation processes. In *Proceedings of the Human Factors and Ergonomics Society 50th Annual Meeting*, Hilton San Francisco Hotel, San Francisco, California, October 16–20 (Vol. 50, pp. 1044–1048).

Erez, M. (1977). Feedback: A necessary condition for the goal setting – Performance relationship. *Journal of Applied Psychology, 62*, 624–627.

Ericsson, E. (2001). Independent driving pattern factors and their influence on fuel-use and exhaust emission factors. *Transportation Research Part D: Transport and Environment, 6*, 325–345.

Ericsson, K. A., & Simon, H. A. (1980). Verbal reports as data. *Psychological Review, 87*, 215–251.

Ericsson, K. A., & Simon, H. A. (1993). *Protocol analysis: Verbal reports as data.* Cambridge, MA: MIT Press.

Ernst, N. A., Jamieson, G. A., & Mylopoulos, J. (2006). Integrating requirements engineering and cognitive work analysis: A case study. In *Presented at the Fourth Conference on Systems Engineering Research.* Los Angeles, CA.

European Commission. (2015). Reducing emissions from transport. *Climate Action.* Retrieved December 21, 2015, from http://ec.europa.eu/clima/policies/transport/index_en.htm

Evans, L. (1979). Driver behaviour effects on fuel consumption in urban driving. *Human Factors, 21*, 389–398.

Fastrez, P., & Haué, J.-B. (2008). Designing and evaluating driver support systems with the user in mind. *International Journal of Human-Computer Studies, 66*(3), 125–131.

Faul, F., Erdfelder, E., Lang, A. G., & Buchner, A. (2007). G*Power 3: A flexible statistical power analysis program for the social, behavioral, and biomedical sciences. *Behavior Research Methods, 39*, 175–191.

Feldman, L., Maibach, E. W., Roser-Renouf, C., & Leiserowitz, A. (2011). Climate on Cable: The nature and impact of global warming coverage on Fox News, CNN, and MSNBC. *The International Journal of Press/Politics, 17*(1), 3–31.

Fiat. (2010). *Eco-driving uncovered.* Turin, Italy.

Field, A. (2009). *Discovering statistics using SPSS.* London: Sage.

Fishbein, M., & Ajzen, I. (1975). *Belief, attitude, intention and behaviour: An introduction to theory and research.* Reading, MA: Addison-Wesley.

Flamm, B. J. (2006). *Environmental, environmental attitudes, and vehicle ownership and use.* University of California, California.

Flemming, S. A. C., Hilliard, A., & Jamieson, G. A. (2008). The need for human factors in the sustainability domain. In *Proceedings of the Human Factors and Ergonomics Society 52nd Annual Meeting*, New York Marriott Marquis Times Square, September 22–26 (pp. 748–752).

Fogg, B. J. (2003). *Persuasive technology: Using computers to change what we think and do.* New York: ACM Press.

Ford. (2013). *Driving to lower fuel consumption – And emissions.* Retrieved January 20, 2016, from http://www.ford.co.uk/OwnerServices/FuelEconomyandEnvironmentalProtection/FuelEfficientEcoDrvingTips

Forsyth, B. A. C., & MacLean, K. E. (2006). Predictive haptic guidance: Intelligent user assistance for the control of dynamic tasks. *IEEE Transactions on Visualization and Computer Graphics, 12*(1), 103–13.

Forta, N. G., Griffin, M. J., & Morioka, M. (2011). Difference thresholds for vibration of the foot: Dependence on frequency and magnitude of vibration. *Journal of Sound and Vibration, 330*(4), 805–815.

Franke, T., Arend, M. G., McIlroy, R. C., & Stanton, N. A. (2016). Ecodriving in hybrid electric vehicles – Exploring challenges for user-energy interaction. *Applied Ergonomics, 55*, 33–45.

Franke, T., Neumann, I., Bühler, F., Cocron, P., & Krems, J. F. (2011). Experiencing Range in an electric vehicle: Understanding psychological barriers. *Applied Psychology: An International Review, 61*(3), 368–391.

Fricke, N., & Schießl, C. (2010). Encouraging environmentally friendly driving through driver assistance: The ecomove project. In *Proceedings of the Sixth International Driving Symposium on Human Factors in Driver Assessment, Training and Vehicle Design*, Lake Tahoe, California, June 27–30 (pp. 394–400).

Fuj, E. T., Hennessy, M., & Mak, J. (1985). An evaluation of the validity and reliability of survey response data on household electricity conservation. *Evaluation Review, 9*, 93–104.

Furukawa, H. (2009). Usage of different levels of functional information in multiple robot operation. In *4th International Conference on Autonomous Robots and Agents*, 10 Feb–12 Feb (pp. 74–78). Wellington, New Zealand: IEEE.

Furukawa, H. (2010). Adaptable user interface based on the ecological interface design concept for multiple robots operating works with uncertainty. *Journal of Computer Science, 6*, 904–911.

Gacias, B., Cegarra, J., & Lopez, P. (2009). An interdisciplinary method for a generic vehicle routing problem decision support system. In *International Conference on Industrial Engineering and Systems Management.* Montréal, Canada.

Gacias, B., Cegarra, J., & Lopez, P. (2010). Work domain analysis and ecological interface for the vehicle routing problem. In *11th IFAC/IFIP/IFORS/IEA Symposium on Analysis, Design and Evaluation of Human- Machine System.* August 31–September 3, Valenciennes, France.

Garnaut, R. (2011). *The garnaut review 2011.* Cambridge: Cambridge University Press.

Gentner, D., & Stevens, A. L. (1983). *Mental models.* Boston, MA: Houghton Mifflin.

Gibson, J. J. (1979). *The ecological approach to visual perception*. Boston, MA: Houghton Mifflin.

Gibson, J. J., & Crooks, L. E. (1938). A theoretical field-analysis of automobile driving. *American Journal of Psychology, 51*, 453–471.

Glaser, R. (1984). Education and thinking: The role of knowledge. *American Psychologist, 39*, 93–104.

Gonder, J., Earleywine, M., & Sparks, W. (2011). *Final report on the fuel saving effectiveness of various driver feedback approaches*. National Renewable Energy Laboratory, Golden, CO.

Gov.uk. (2014). *General rules, techniques and advice for all drivers and riders (103 to 158)*. Retrieved March 20, 2014, from https://www.gov.uk/general-rules-all-drivers-riders-103-to-158/control-of-the-vehicle-117-to-126

Groeger, J. A. (2000). *Understanding driving: Applying cognitive psychology to a complex everyday task*. Hove: Psychology Press.

Groskamp, P. A., van Paassen, M. M., & Mulder, M. (2005). Interface design for engagement planning in anti-air warfare. In *IEEE International Conference on Systems, Man, and Cybernetics*, 12 Oct, Waikoloa, HI (pp. 311–316).

Gu, C., & Griffin, M. J. (2012). Vibrotactile perception thresholds at the sole of the foot: Effects of contact force and probe indentation. *Medical Engineering and Physics, 34*(4), 447–452.

Hajdukiewicz, J. R., Vicente, K. J., Doyle, D. J., Milgram, P., & Burns, C. M. (2001). Modeling a medical environment: An ontology for integrated medical informatics design. *International Journal of Medical Informatics, 62*(1), 79–99.

Hajek, H., Popiv, D., Just, M., & Bengler, K. (2011). Influence of a multimodal assistance supporting anticipatory driving on the driving behavior and driver's acceptance. In M. Kurosu (Ed.), *Human centered design, HCII 2011, LNCS 6776* (pp. 217–226). Berlin: Springer-Verlag.

Hall, D. S., Shattuck, L. G., & Bennett, K. B. (2012). Evaluation of an ecological interface design for military command and control. *Journal of Cognitive Engineering and Decision Making, 6*, 165–193.

Ham, D., & Yoon, W. C. (2001a). Design of information content and layout for process control based on goal-means domain analysis. *Cognition, Technology & Work, 3*, 205–223.

Ham, D., & Yoon, W. C. (2001b). The effects of presenting functionally abstracted information in fault diagnosis tasks. *Reliability Engineering & System Safety, 73*(2), 103–119.

Hansen, J. P. (1995). Representation of system invariants by optical invariants in configural displays for process control. In P. Hancock, J. Flach, J. Caird, & K. Vicente (Eds.), *Local applications of the ecological approach to human-machine systems* (pp. 208–233). Hillsdale, NJ: Lawrence Erlbaum Associates.

Harms, L., & Patten, C. (2003). Peripheral detection as a measure of driver distraction. A study of memory-based versus system-based navigation in a built-up area. *Transportation Research Part F: Traffic Psychology and Behaviour, 6*(1), 23–36.

Hart, S. G. (2006). NASA-Task Load Index (NASA-TLX); 20 years later. In *Proceedings of the Human Factors and Ergonomics Society 50th Annual Meeting* (Vol. 50, pp. 904–908). Sage.

Hart, S. G., & Staveland, L. E. (1988). Development of NASA-TLX (Task Load Index): Results of empirical and theoretical research. In P. A. Hancock & N. Meshkati (Eds.), *Human mental workload* (pp. 139–183). Amsterdam: Elsevier Science.

Harvey, C., & Stanton, N. A. (2012). Trade-off between context and objectivity in an analytic approach to the evaluation of in-vehicle interfaces. *IET Intelligent Transport Systems, 6*, 243–258.

Harvey, C., & Stanton, N. A. (2013). *Usability evaluation for in-vehicle systems*. England: CRC Press.

Harvey, C., Stanton, N. A., Pickering, C. A., McDonald, M., & Zheng, P. (2011a). A usability evaluation toolkit for In-vehicle Information Systems (IVISs). *Applied Ergonomics, 42*(4), 563–574.

Harvey, C., Stanton, N. A., Pickering, C. A., McDonald, M., & Zheng, P. (2011b). In-vehicle information systems to meet the needs of drivers. *International Journal of Human-Computer Interaction, 27*(6), 505–522.

Harvey, J., Thorpe, N., & Fairchild, R. (2013). Attitudes towards and perceptions of eco-driving and the role of feedback systems. *Ergonomics, 56*(3), 37–41.

Hawkins, T. R., Singh, B., Majeau-Bettez, G., & Strømman, A. H. (2013). Comparative environmental life cycle assessment of conventional and electric vehicles. *Journal of Industrial Ecology, 17*, 53–64.

Haworth, N., & Symmons, M. (2001). *The relationship between fuel economy and safety outcomes*. MONASH, Clayton, Victoria, Australia.

Hayes, S. C. (1986). The case of the silent dog – Verbal reports and the analysis of rules: A review of Ericsson and Simon's protocol analysis: Verbal reports as data 1. *Journal of the Experimental Analysis of Behavior, 45*(3), 351–363.

Hedges, P., & Moss, D. (1996). Costing the effectiveness of training: Case study 1-improving parcelforce driver performance. *Industrial and Commercial Training, 28*, 14–18.

Hibberd, D. L., Jamson, H., & Jamson, S. L. (2015). The design of an in-vehicle assistance system to support eco-driving. *Transportation Research Part C: Emerging Technologies, 58*, 732–748.

Hill, N., Brannigan, C., Smokers, R., Skinner, I., Schroten, A., & Van Essen, H. (2012). *EU Transport GHG: Routes to 2050 II developing a better understanding of the secondary impacts and key sensitivities for the decarbonisation of the EU's transport sector by 2050. Final project report produced as part of a contract between European Commission Directorate-General Climate Action and AEA Technology plc.* http://www.eutransportghg2050.eu/cms/assets/Uploads/Reports/EU-Transport-GHG-2050-II-Final-Report-29Jul12.pdf (Accessed December 15, 2015).

Hilliard, A., & Jamieson, G. A. (2007). Ecological interface design for solar car strategy: From state equations to visual relations. In *2007 IEEE International Conference on Systems, Man and Cybernetics*, Montreal, Quebec, Canada, 7–10 Oct (pp. 139–144).

Hilliard, A., & Jamieson, G. A. (2008). Winning the solar races with interface design. *Ergonomics in Design, 16*, 6–11.

Hilliard, A., & Jamieson, G. A. (2014). A strategy-based ecological (?) display for time-series structural change diagnosis. In *2014 IEEE International Conference on Systems, Man, and Cybernetics* (pp. 353–359).

Hines, J. M., Hungerford, H. R., & Tomera, A. N. (1987). Analysis and synthesis of research on responsible environmental behavior: A meta-analysis. *The Journal of Environmental Education, 18*(2), 1–8.

Hmielowski, J. D., Feldman, L., Myers, T. A., Leiserowitz, A., & Maibach, E. (2014). An attack on science? Media use, trust in scientists, and perceptions of global warming. *Public Understanding of Science, 23*(7), 866–883.

Ho, C., & Spence, C. (2008). *The multisensory driver: Implications for ergonomic car interface*. Aldershot: Ashgate Publishing Limited.

Ho, C., Tan, H. Z., & Spence, C. (2006). The differential effect of vibrotactile and auditory cues on visual spatial attention. *Ergonomics, 49*(7), 724–738.

Holmén, B. A., & Niemeier, D. A. (1998). Characterizing the effects of driver variability on real-world vehicle emissions. *Transportation Research Part D: Transport and the Environment, 3*, 117–128.

Hooker, J. N. (1988). Optimal driving for single-vehicle fuel economy. *Transportation Research Part A: Policy and Practice, 22*(3), 183–201.

Horberry, T., Anderson, J., Regan, M. A., Triggs, T. J., & Brown, J. (2006). Driver distraction: The effects of concurrent in-vehicle tasks, road environment complexity and age on driving performance. *Accident; Analysis and Prevention, 38*(1), 185–191.

Horberry, T., Stevens, A., Burnett, G., Cotter, S., & Robbins, R. (2008). Assessing the visual demand from in-vehicle information systems by means of the occlusion technique: The effects of participant age. *IET Intelligent Transport Systems, 2,* 170–177.

Horiguchi, Y., Asakura, R., Sawaragi, T., Tamai, Y., Naito, K., Hashiguchi, N., & Konishi, H. (2007). Ecological interface to enhance user performance in adjusting computer-controlled multihead weigher. In M. J. Smith & G. Salvendy (Eds.), *Human interface part II, HCII 2007* (pp. 883–892). Berlin: Springer-Verlag.

Horiguchi, Y., Kurono, K., Nakanishi, H., Sawaragi, T., Nagatani, T., Noda, A., & Tanaka, K. (2010). Ecological interface design for teaching assembly operations to industrial robot. In *11th IFAC,IFIP,IFORS, IEA Symposium on Analysis, Design, and Evaluation of Human-Machine Systems*, Aug 31–Sep 03, Valenciennes, France (Vol. 11, pp. 442–447).

Hotelling, H. (1931). The generalization of student's ratio. *The Annals of Mathematical Statistics, 2,* 360–378.

Howie, D. E., & Vicente, K. J. (1998a). Making the most of ecological interface design: The role of self-explanation. *International Journal of Human-Computer Studies, 49,* 651–674.

Howie, D. E., & Vicente, K. J. (1998b). Measures of operator performance in complex, dynamic microworlds: Advancing the state of the art. *Ergonomics, 41*(4), 485–500.

Hughes, K. P., & Cole, B. L. (1986). What attracts attention when driving? *Ergonomics, 29*(3), 377–391.

Hulme, M., & Turnpenny, J. (2004). Understanding and managing climate change: The UK experience. *The Geographical Journal, 170*(2), 105–115.

Hutchins, E. L., Hollan, J. D., & Norman, D. A. (1986). Direct manipulation interfaces. In D. A. Norman & S. W. Draper (Eds.), *User centered system design: New perspectives on human-computer interaction* (pp. 87–124). Hillsdale, NJ: Lawrence Erlbaum Associates.

Intergovernmental Panel on Climate Change. (2007). *Climate change 2007: Synthesis report.* Cambridge, UK: Cambridge University Press.

International Energy Authority. (2012). *World energy outlook.* Paris: OECD.

IPCC. (2014). *Climate change 2014: Synthesis Report: Contribution of Working Groups I, II and III to the Fifth Assessment Report of the Intergovernmental Panel on Climate Change.* R. K. Pachauri & L. A. Meyer (Eds.). Geneva, Switzerland.

Isenberg, D. J. (1986). Thinking and managing: A verbal protocol analysis of managerial problem solving. *Academy of Management Journal, 29,* 775–788.

Itoh, J., Sakuma, A., & Monta, K. (1995). An ecological interface design for supervisory control of BWR nuclear power plants. *Control Engineering Practice, 3*(2), 231–239.

Jackson, J. S. H., & Blackman, R. (1994). A driving simulator test of Wilde's risk homeostasis theory. *Journal of Applied Psychology, 79,* 950–958.

Jackson, T. (2005). *Motivating sustainable consumption: A review of evidence on consumer behaviour and behavioural change. Report to the Sustainable Development Research Network.* Centre for Environmental Strategy, Guildford, Surry, UK.

Jager, W. (2003). Breaking 'bad habits': A dynamical perspective on habit formation and change. In L. Hendrickx, W. Jager, & L. Steg (Eds.), *Human decision making and environmental perception. Understanding and assisting human decision making in real-life settings.* Groningen: University of Groningen.

Jamieson, G. A. (2002). Empirical evaluation of an industrial application of ecological interface design. In *Proceedings of the 46th Annual Meeting of the Human Factors and Ergonomics Society*, Baltimore Marriott Waterfront Hotel, Baltimore, Maryland, September 30–October 4 (pp. 536–540). Santa Monica, CA: Human Factors and Ergonomics Society.

Jamieson, G. A., & Ho, W. H. (2001). *A prototype ecological interface for a simulated petro-chemical process* (Vol. 1). Cognitive Engineering Laboratory, Toronto, ON.

Jamieson, G. A., Miller, C. A., Ho, W. H., & Vicente, K. J. (2007). Integrating task- and work domain-based work analyses in ecological interface design: A process control case study. *IEEE Transactions on Systems, Man, and Cybernetics – Part A: Systems and Humans, 37*, 887–905.

Jamieson, G. A., Reising, D. V. C., & Hajdukiewicz, J. (2001). *EID design rationale project: Case study report* (Vol. 1). Cognitive Engineering Laboratory, Toronto, ON.

Jamieson, G. A., & Vicente, K. J. (2001). Ecological interface design for petrochemical applications: Supporting operator adaptation, continuous learning, and distributed, collaborative work. *Computers & Chemical Engineering, 25*(7–8), 1055–1074.

Jamson, A. H., Hibberd, D. L., & Merat, N. (2015). Interface design considerations for an in-vehicle eco-driving assistance system. *Transportation Research Part C, 58*, 642–656.

Jamson, H., Hibberd, D. L., & Merat, N. (2013). The design of haptic gas pedal feedback to support eco-driving. In *Proceedings of the Seventh International Driving Symposium on Human Factors in Driver Assessment, Training and Vehicle Design*, June 17–20, Bolton Landing, New York (pp. 264–270).

Janssen, W., & Nilsson, L. (1993). Behavioural effects of driver support. In A. M. Parkes & S. Franzen (Eds.), *Driving future vehicles* (pp. 147–155). Washington, DC: Taylor and Francis.

Jenkins, D. P., Stanton, N. A., Salmon, P. M., & Walker, G. H. (2009). *Cognitive work analysis: Coping with complexity*. Farnham, England: Ashgate Publishing Limited.

Jenkins, D. P., Stanton, N. A., Salmon, P. M., Walker, G. H., & Rafferty, L. (2010). Using the decision ladder to add a formative element to naturalistic decision-making research. *International Journal of Human-Computer Interaction, 26*(2–3), 132–146.

Jenkins, D. P., Stanton, N. A., Walker, G. H., & Young, M. S. (2007). A new approach to designing lateral collision warning systems. *International Journal of Vehicle Design, 45*(3), 379–396.

Jensen, B. B. (2002). Knowledge, action and pro-environmental behaviour. *Environmental Education Research, 8*(3), 325–334.

Jipp, M., Schaper, M., Guenther, Y., & Papenfuss, A. (2011). Ecological interface design and its application to total airport management. In *2011 IEEE International Conference on Systems, Man, and Cybernetics*, 9–12 Oct, Anchorage, AK (pp. 2101–2106). IEEE.

Jungk, A., Thull, B., Hoeft, A., & Rau, G. (1999). Ergonomic evaluation of an ecological interface and a profilogram display for hemodynamic monitoring. *Journal of Clinical Monitoring and Computing, 15*(7–8), 469–479.

Jungk, A., Thull, B., Hoeft, A., & Rau, G. (2000). Evaluation of two new ecological interface approaches for the anesthesia workplace. *Journal of Clinical Monitoring and Computing, 16*(4), 243–258.

Kaufmann, C., Risser, R., Geven, A., Sefelin, R., & Tschelig, M. (2008). LIVES (LenkerInnenInteraktion mit VErkehrstelematischen Systemen) – Driver interaction with transport-telematic systems. *IET Intelligent Transport Systems, 2*(4), 294–305.

Kidd, D. G. (2012). Response of part-time belt users to enhanced seat belt reminder systems of different duty cycles and duration. *Transportation Research Part F: Traffic Psychology and Behaviour, 15*(5), 525–534.

Kim, S., Dandekar, H., Camez, E., & Harrelson, H. (2011). Assessing the effect of a power-flow gauge on driving behaviors affecting energy consumption. In J. A. Jacko (Ed.), *Human-computer interaction, part III, HCII 2011, LNCS 6763* (pp. 411–417). Berlin: Springer-Verlag.

Kim, S. K., Suh, S. M., Jang, G. S., Hong, S. K., & Park, J. C. (2012). Empirical research on an ecological interface design for improving situation awareness of operators in an advanced control room. *Nuclear Engineering and Design, 253*, 226–237.

Kim, S. Y., & Kim, Y. S. (2012). *A virtual driving simulator for eco-driving training.* *6*(2), 358–360. http://onlinepresent.org/proceedings/vol2_2012/112.pdf (Accessed November 23, 2015).

King, J. (2007). *The king review of low-carbon cars. Part I: The potential for CO₂ reduction.* The Stationary Office, London.

King, P. (2011). *AA member eco-driver survey.* AA Research Foundation.

Klomp, R., Borst, C., Mulder, M., & Praetorius, G. (2014). Ecological interface design: Control space robustness in future trajectory-based air traffic control decision support. *In 2014 IEEE International Conference on Systems, Man, and Cybernetics* (pp. 335–340). 5–8 October, San Diego, CA.

Klomp, R., Borst, C., Mulder, M., Praetorius, G., Mooij, M., & Nieuwenhuisen, D. (2013). Experimental evaluation of a joint cognitive system for 4D trajectory management. In D. Schaefer (Ed.), *Proceedings of the Third SESAR Innovation Days.* 26–28 November, Stockholm, Sweden.

Klomp, R., Van Paassen, M. M. R., Borst, C., & Mulder, M. (2012). Joint human-automation cognition through a shared representation of 4D trajectory management. In D. Schaefer (Ed.*), Proceedings of the Second SESAR Innovation Days.* November 27–29, Braunshweig, Germany.

Kluger, A. N., & DeNisi, A. (1996). The effects of feedback interventions on performance: A historical review, a meta-analysis, and a preliminary feedback intervention theory. *Psychological Bulletin, 119*(2), 254–284.

Knapton, S. (2014). BBC staff told to stop inviting cranks on to science programmes. *The Telegraph.* Retrieved January 15, 2016, from http://www.telegraph.co.uk/culture/tvandradio/bbc/10944629/BBC-staff-told-to-stop-inviting-cranks-on-to-science-programmes.html

Kollmuss, A., & Agyeman, J. (2002). Mind the gap: Why do people act environmentally and what are the barriers to pro-environmental behavior? *Environmental Education Research, 8*(3), 239–260.

Krems, J., Franke, T., Neumann, I., & Cocron, P. (2010). Research methods to assess the acceptance of EVs – Experiences from an EV user study. In T. Gessner (*Ed.), Smart Systems Integration: 4th European Conference & Exhibition on Integration Issues of Miniaturized Systems – MEMS, MOEMS, ICs and Electronic Components.* 23–24 March, Como, Italy: VDE Verlag.

Kruit, J. D., Amelink, M., Mulder, M., & van Paassen, M. M. (2005). Design of a rally driver support system using ecological interface design principles. In *2005 IEEE International Conference on Systems, Man and Cybernetics* 10–12 October, Waikoloa, HI (Vol. 2, pp. 1235–1239).

Kujala, T., & Saariluoma, P. (2011). Effects of menu structure and touch screen scrolling style on the variability of glance durations during in-vehicle visual search tasks. *Ergonomics, 54*(8), 716–732.

Kuusela, H., & Paul, P. (2000). A comparison of concurrent and retrospective verbal protocol analysis. *American Journal of Psychology, 113*(3), 387–404.

Kwok, J., & Burns, C. M. (2005). Usability evaluation of a mobile ecological interface design application for diabetes management. In *Proceedings of the Human Factors and Ergonomics Society 49th Annual Meeting*, Loews Royal Pacific Resort, Orlando, Florida, September 26–30 (Vol. 49, pp. 1042–1046).

Lajunen, T., & Summala, H. (2003). Can we trust self-reports of driving? Effects of impression management on driver behaviour questionnaire responses. *Transportation Research Part F: Traffic Psychology and Behaviour, 6*(2), 97–107.

Landauer, T. K. (1997). Behavioural research methods in human-computer interaction. In M. Helander, T. K. Landauer, & P. Prabhu (Eds.), *Handbook of human-computer interaction* (pp. 203–227). Oxford: Elsevier Science.

Lansdown, T. C. (2002). Individual differences during driver secondary task performance: Verbal protocol and visual allocation findings. *Accident Analysis and Prevention, 34*(5), 655–662.

Lansdown, T. C., Brook-Carter, N., & Kersloot, T. (2004). Distraction from multiple in-vehicle secondary tasks: Vehicle performance and mental workload implications. *Ergonomics, 47*(1), 91–104.

Larsson, H., & Ericsson, E. (2009). The effects of an acceleration advisory tool in vehicles for reduced fuel consumption and emissions. *Transportation Research Part D: Transport and Environment, 14*(2), 141–146.

Lau, N., & Jamieson, G. A. (2006). Ecological interface design for the condenser subsystems of a boiling water reactor. In *Proceedings of the 27th Annual Conference of the Canadian Nuclear Society*. Toronto, ON: Nuclear Society.

Lau, N., Jamieson, G. A., Skraaning, G., & Burns, C. M. (2008). Ecological interface design in the nuclear domain: An empirical evaluation of ecological displays for the secondary subsystems of a boiling water reactor plant simulator. *IEEE Transactions on Nuclear Science, 55*(6), 3597–3610.

Lau, N., Veland, Ø., Kwok, J., Jamieson, G. A., Burns, C. M., Braseth, A. O., & Welch, R. (2008). Ecological interface design in the nuclear domain: An application to the secondary subsystems of a boiling water reactor plant simulator. *IEEE Transactions on Nuclear Science, 55*(6), 3579–3596.

Lee, J. D., Hoffman, J. D., Stoner, H. A., Seppelt, B. D., & Brown, M. D. (2006). Application of ecological interface design to driver support systems. *In Proceedings of IEA 2006: 16th World Congress on Ergonomics*. Maastricht, The Netherlands.

Lee, J. D., Stoner, H. A., & Marshall, D. (2004). Enhancing interaction with the driving ecology through haptic interfaces. 2004 *IEEE International Conference on Systems, Man and Cybernetics (IEEE Cat. No.04CH37583)* (Vol. 1, pp. 841–846).

Lee, S. (2012). Design of the cognitive information display for water level control of the steam generator in Korean nuclear power reactor. *Network and Parallel Computing Lecture Notes in Computer Science, 7513*, 636–644.

Lee, S. W., Nam, T. S., & Myung, R. (2008). Work Domain Analysis (WDA) for Ecological Interface Design (EID) of vehicle control display. In *9th WSEAS International Conference on AUTOMATION and INFORMATION (ICAI'08)*, June 24–26 (pp. 387–392). Bucharest, Romania.

Lehane, P., Toleman, M., & Benecke, J. (2000). Applying ecological interface design to experimental apparatus used to monitor a refrigeration plant. In *Proceedings of First Australian User Interface Conference*, January 31–February 3, Canberra, Australia (pp. 41–48).

Li, Y., Burns, C. M., & Kuli, D. (2014). Ecological interface design for knee and hip automatic physiotherapy assistant and rehabilitation system. *Proceedings of the International Symposium of Human Factors and Ergonomics in Healthcare, 3*(1), 1–7.

Lichtenstein, S., & Fischhoff, B. (1977). Do those who know more also know more about how much they know? *Organizational Behavior and Human Performance, 183*(3052), 159–183.

Lin, H., Cui, X., Yu, Q., & Yang, S. (2011). Experimental study on diesel vehicle's fuel consumption feature while coasting on level road. In *CSIE, Part I, CCIS 152* (pp. 264–270). Berlin: Springer-Verlag.

Lindgren, A., Angelelli, A., Mendoza, P. A., & Chen, F. (2009). Driver behaviour when using an integrated advisory warning display for advanced driver assistance systems. *IET Intelligent Transport Systems, 3*(4), 390–399.

Linegang, M. P., Stoner, H. A., Patterson, M. J., Seppelt, B. D., Hoffman, J. D., Crittendon, Z. B., & Lee, J. D. (2006). Human-automation collaboration in dynamic mission planning: A challenge requiring a ecological approach. In *Proceedings of the Human Factors and Ergonomics Society 50th Annual Meeting*, Hilton San Francisco Hotel, San Francisco, California, October 16–20 (pp. 2482–2486).

Lintern, G., Miller, D., & Baker, K. (2002). Work centered design of a Usaf mission planning system. In *Proceedings of the Human Factors and Ergonomics Society 46th Annual Meeting*, Baltimore Marriott Waterfront Hotel, Baltimore, Maryland, September 30–October 4 (Vol. 46, pp. 531–535).

Liu, Q., Itoh, H., & Yoshimura, K. (2006). On the design of in-vehicle advice system. In *2006 IEEE International Conference on Systems, Man, and Cybernetics*, 8–11 October (pp. 2511–2516). Taipei, Taiwan.

Locke, E. A. (1967). The motivational effects of knowledge of results: Knowledge or goal setting? *Journal of Applied Psychology, 51*, 324–329.

Locke, E. A. (1968). Toward a theory of task motivation and incentives. *Organizational Behavior and Human Performance, 3*, 157–189.

Locke, E. A., & Bryan, J. F. (1968). Goal setting as a determinant of the effect of knowledge of score on performance. *American Journal of Psychology, 81*, 398–406.

Locke, E. A., & Bryan, J. F. (1969a). Knowledge of score and goal level as determinants of work rate. *Journal of Applied Psychology, 53*, 59–65.

Locke, E. A., & Bryan, J. F. (1969b). The directing function of goals in task performance. *Organizational Behavior and Human Performance, 4*, 35–42.

Locke, E. A., & Latham, G. P. (2002). Building a practically useful theory of goal setting and task motivation: A 35-year odyssey. *American Psychologist, 57*(9), 705–717.

Lockton, D., Harrison, D., & Stanton, N. A. (2008a). Design with intent: Persuasive technology in a wider context. In H. Oinas-Kukkonen (Ed.), *Persuasive 2008, LNCS 5033* (pp. 274–278). Berlin: Springer-Verlag.

Lockton, D., Harrison, D., & Stanton, N. A. (2008b). Making the user more efficient: Design for sustainable behaviour. *International Journal of Sustainable Engineering, 1*(1), 3–8.

Lockton, D., Harrison, D., & Stanton, N. A. (2010). The design with intent method: A design tool for influencing user behaviour. *Applied Ergonomics, 41*(3), 382–392.

Marshall, D. C., Lee, J. D., & Austria, R. A. (2007). Alerts for in-vehicle information systems: Annoyance, urgency, and appropriateness. *Human Factors, 49*(1), 145–157.

Martin, K., Legg, S., & Brown, C (2012). Designing for sustainability: Ergonomics – Carpe diem. *Ergonomics*, September, pp. 37–41.

Matosin, N., Frank, E., Engel, M., Lum, J. S., & Newell, K. A (2014). Negativity towards negative results: A discussion of the disconnect between scientific worth and scientific culture. *Disease Models and Mechanisms, 7*, 171–173.

Mayer, R. E. (2001). *Multimedia learning*. New York: Cambridge University Press.

McCalley, L. T. (2006). From motivation and cognition theories to everyday applications and back again: The case of product-integrated information and feedback. *Energy Policy, 34*(2), 129–137.

McCalley, L. T., & Midden, C. J. (2002). Energy conservation through product-integrated feedback: The roles of goal-setting and social orientation. *Journal of Economic Psychology, 23*(5), 589–603.

McCright, A. M., Dunlap, R. E., & Xiao, C. (2013). Perceived scientific agreement and support for government action on climate change in the USA. *Climatic Change, 119*(2), 511–518.

McEwen, T., Flach, J., & Elder, N.(2012). Ecological interface design for assessing cardiac disease. In *Proceedings of the ASME 2012 11th Biennial Conference On Engineering Systems Design And Analysis ESDA2012*. July 2–4, pp. 881–888. Nantes, France.

McEwen, T. R., Flach, J. M., & Elder, N. C. (2014). Interfaces to medical information systems: Supporting evidenced based practice. In *2014 IEEE International Conference on Systems, Man, and Cybernetics*, October 5–8 (pp. 341–346). San Diego, CA.

McIlroy, R. C., & Stanton, N. A. (2011). Getting past first base: Going all the way with cognitive work analysis. *Applied Ergonomics, 42*(2), 358–370.

McIlroy, R. C., & Stanton, N. A. (2012). Specifying the requirements for requirements specification: The case for work domain and worker competencies analyses. *Theoretical
Issues in Ergonomics Science, 13*(4), 450–471. doi:10.1080/1463922X.2010.539287

McIlroy, R. C., Stanton, N. A., & Remington, B. (2012). Developing expertise in military
communications planning: Do verbal reports change with experience? *Behaviour &
Information Technology, 13*, 37–41.

McKenna, F. P., Stanier, R. A., & Lewis, C. (1991). Factors underlying illusory self- assessment of driving skill in males and females. *Accident Analysis and Prevention, 23*(1),
45–52.

Memisevic, R., Sanderson, P., Choudhury, S., & Wong, B. L. W. (2005). Work domain analysis and ecological interface design for hydropower system monitoring and control. In
2005 IEEE International Conference on Systems, Man and Cybernetics (Vol. 4).

Mendoza, P., Angelelli, A., & Lindgren, A. (2011). Ecological interface design inspired human
machine interface for advanced driver assistance systems. *IET Intelligent Transport
Systems, 5*(1), 53–59.

Milfont, T. L., & Duckitt, J. (2010). The environmental attitudes inventory: A valid and reliable
measure to assess the structure of environmental attitudes. *Journal of Environmental
Psychology, 30*(1), 80–94.

Miller, C. A., & Vicente, K. J. (1998). Toward an integration of task- and work domain anal-
ysis techniques for human-computer interface design. In *Proceedings of the Human
Factors and Ergonomics Society Annual 42nd Meeting*, Hyatt Regency Chicago,
Chicago, Illinois, October 5–9 (Vol. 42, pp. 336–340).

Mitsopoulos-Rubens, E., Trotter, M. J., & Lenné, M. G. (2011). Usability evaluation as part of
iterative design of an in-vehicle information system. *IET Intelligent Transport Systems,
5*(2), 112.

Monta, K., Ohtsu, M., & Inagaki, T. (1999). A study on effectiveness of automation and
interface presentation for coping with transients. In *IEEE International Conference on
Systems, Man, and Cybernetics*, October 12–15 (pp. 751–756). Tokyo, Japan.

Moore, D. A., & Small, D. A. (2007). Error and bias in comparative judgment: On being both
better and worse than we think we are. *Journal of Personality and Social Psychology,
92*(6), 972–989.

Morineau, T., Beuzet, E., Rachinel, A., & Tobin, L. (2009). Experimental evaluation of a
tide prediction display based on the ecological interface design framework. *Cognition,
Technology & Work, 11*, 119–127.

Mulder, M. (2007). *Haptic gas pedal feedback for active car-following support*. Delft
University of Technology. Delft, The Netherlands.

Mulder, M., Abbink, D. A., van Paassen, M. M., & Mulder, M. (2011). Design of a haptic gas
pedal for active car-following support. *IEEE Transactions on Intelligent Transportation
Systems, 12*(1), 268–279.

Mulder, M., Pauwelussen, J. J. A., van Paassen, M. M., & Abbink, D. A. (2010). Active deceleration support in car following. *IEEE Transactions on Systems, Man, and Cybernetics
– Part A: Systems and Humans, 40*, 1271–1284.

Mulder, M., van Paassen, M. M., & Abbink, D. A. (2008). Haptic gas pedal feedback.
Ergonomics, 51(11), 1710–1720.

Muñoz-Organero, M., & Magaña, V. C. (2013). Validating the impact on reducing fuel consumption by using an ecodriving assistant based on traffic sign detection and optimal
deceleration patterns. *IEEE Transactions on Intelligent Transportation Systems, 14*(2),
1023–1028.

Naikar, N. (2006). An examination of the key concepts of the five phases of cognitive work
analysis with examples from a familiar situation. In *Proceedings of the Human Factors
and Ergonomics Society 50th Annual Meeting*, Hilton San Francisco Hotel, San
Francisco, California, October 16–20 (pp. 447–451).

Naikar, N., Moylan, A., & Pearce, B. (2006). Analysing activity in complex systems with cognitive work analysis: Concepts, guidelines and case study for control task analysis. *Theoretical Issues in Ergonomics Science, 7*(4), 371–394.

Naito, N., Itoh, J., Monta, K., & Makino, M. (1995). An intelligent human-machine system based on an ecological interface design concept. *Nuclear Engineering and Design, 154,* 97–108.

Nam, T. S., & Myung, R. (2007). The application of work domain analysis (WDA) for the development of vehicle control display. In *Proceedings of the 7th WSEAS International Conference on Applied Informatics and Communications* (pp. 160–165). Athens, Greece.

Neisser, U. (1976). *Cognition and reality.* San Francisco, CA: W.H.Freemand & Co.

Ngo, M. K., & Spence, C. (2010). Auditory, tactile, and multisensory cues facilitate search for dynamic visual stimuli. *Attention, Perception, & Psychophysics, 72*(6), 1654–1665.

Nisbett, R. E., & Wilson, T. D. (1977). Telling more than we can know: Verbal reports on mental processes. *Psychological Review, 84,* 231–259.

Lewis, C. & Norman, D.A. (1986) Designing for error. In User centered system design, D.A. Norman & S.W. Draper (Eds), 411–432. Hillsdale, NJ: Lawrence Erlbaum Associates.

Nunnally, J. C. (1978). *Psychometric theory.* New York: McGraw-Hill.

Nykvist, B., & Nilsson, M. (2015). Rapidly falling costs of battery packs for electric vehicles. *Nature Climate Change, 5,* 329–332.

O'Connell, S. (1998). *The car and British society: Class, gender and motoring, 1896–1939.* Manchester, UK: Manchester University Press.

Office for National Statistics. (2012). *Measuring national well-being, education and skills.* The Stationary Office, London.

Olsheski, J. D. (2012). Multimodal display integration for in-vehicle assistive technology (IVAT). SIG *Access Newsletter, 102,* 30–35.

Olsson, G., & Lee, P. L. (1994). Effective interfaces for process operators – A prototype. *Journal of Process Control, 4*(2), 99–107.

Organisation of Economic Cooperation and Development. (1993). *Cars and climate change.* Paris: OECD.

Pampel, S. M., Jamson, S. L., Hibberd, D. L., & Barnard, Y. (2015). How I reduce fuel consumption: An experimental study on mental models of eco-driving. *Transportation Research Part C: Emerging Technologies, 58,* 669–680.

Pawlak, W. S., & Vicente, K. J (1996). Inducing effective operator control through ecological interface design. *International Journal of Human-Computer Studies, 44*(5), 653–688.

Pearre, N. S., Kempton, W., Guensler, R. L., & Elango, V. V. (2011). Electric vehicles: How much range is required for a day's driving? *Transportation Research Part C: Emerging Technologies, 19*(6), 1171–1184.

Pejtersen, A. M. (1989). *The Bookhouse: Modelling user needs and search strategies a basis for system design.* Roskilde, Denmark: Riso National Laboratory.

Piaget, J. (1952). *The origins of intelligence in children.* New York: International University Press.

Pinder, S. D., Bristow, D. N., & Davies, T. C. (2006). *Interface design for an aircraft thrust and braking indicator/advisor. In OZCHI '06 Proceedings of the 18th Australia conference on Computer-Human Interaction: Design: Activities, Artefacts and Environments,* 103–110 Sydney, Australia, November 20–24.

Pirolli, P., & Recker, M. (1994). Learning strategies and transfer in the domain of programming. *Cognition and Instruction, 12,* 235–275.

Plant, K. L., & Stanton, N. A. (2012). The explanatory power of schema theory: Theoretical foundations and future applications in ergonomics. *Ergonomics,* December, pp. 37–41.

Plant, K. L., & Stanton, N. A. (2015). The process of processing: Exploring the validity of Neisser's perceptual cycle model with accounts from critical decision-making in the cockpit. *Ergonomics, 58*(6), 909–923.

Poortinga, W., Spence, A., Whitmarsh, L., Capstick, S., & Pidgeon, N. F. (2011). Uncertain climate: An investigation into public scepticism about anthropogenic climate change. *Global Environmental Change, 21*(3), 1015–1024.

Pradhan, A. K., Divekar, G., Masserang, K., Romoser, M., Zafian, T., Blomberg, R. D., Thomas, F. D., et al. (2011). The effects of focused attention training on the duration of novice drivers' glances inside the vehicle. *Ergonomics, 54*(10), 917–931.

Precision Microdrives. (2015). *AB-004: Understanding ERM vibration motor characteristics.* Retrieved January 7, 2015, from http://www.precisionmicrodrives.com/application-notes-technical-guides/application-bulletins/ab-004-understanding-erm-characteristics-for-vibration-applications

Precision Microdrives. (2016). *Why is vibration amplitude in G?* Retrieved March 27, 2016, from http://www.precisionmicrodrives.com/tech-blog/2013/02/25/why-is-vibration-amplitude-in-g

Rasmussen, J. (1974). *The human data processor as a system component. Bits and pieces of a model. Riso-M-1722.* Danish Atomic Energy Commission, Roskilde, Denmark.

Rasmussen, J. (1983). Skills, rules, and knowledge; signals, signs, and symbols, and other distinctions in human performance models. *IEEE Transactions on Systems, Man, and Cybernetics, 13*(3), 257–266.

Rasmussen, J. (1985). The role of hierarchical knowledge representation in decisionmaking and system management. *IEEE Transactions on Systems, Man, and Cybernetics, 15*, 234–243.

Rasmussen, J. (1986). *Information processing and human-machine interaction.* Amsterdam: North-Holland.

Rasmussen, J., Pejtersen, A., & Goodstein, L. P. (1994). *Cognitive systems engineering.* New York: Wiley.

Rasmussen, J., Pejtersen, A. M., & Schmidt, K. (1990). *Taxonomy for cognitive work analysis.* Roskilde, Denmark: Risø National Laboratory.

Rasmussen, J., & Vicente, K. J. (1989). Coping with human errors through system design: Implications for ecological interface design. *International Journal of Man-Machine Studies, 31*(5), 517–534.

Rayner, K. (1997). Eye movements in reading and information processing: 20 years of research. *Psychological Bulletin, 124*(3), 372–422.

Read, G. J. M., Salmon, P. M., & Lenné, M. G. (2012). From work analysis to work design: A review of cognitive work analysis design applications. In *Proceedings of the Human Factors and Ergonomics Society 56th Annual Meeting*, Westin Boston Waterfront, Boston, Massachusetts, October 22–26 (pp. 368–372).

Reason, J. (1990). *Human error.* New York: Cambridge University Press.

Redström, J. (2006). Persuasive design: Fringes and foundations. In W. I. Jsselsteijn (Ed.), *Persuasive 2006, LNCS 3962* (pp. 112–122). Berlin: Springer-Verlag.

Reising, D. V. C., & Sanderson, P. M. (2002a). Ecological interface design for Pasteurizer II: A process description of semantic mapping. *Human Factors, 44*, 222–247.

Reising, D. V. C., & Sanderson, P. M. (2002b). Work domain analysis and sensors II: Pasteurizer II case study. *International Journal of Human-Computer Studies, 56*, 597–637.

Renault. (2013). *Eco driving tips.* Retrieved February 27, 2013, from http://www.renault.co.uk/cars/environment/tips.aspx

Reser, J. P., Pidgeon, N., Spence, A., Bradley, G., Glendon, A. I., & Ellul, M. (2011). *Public perceptions, understanding, and responses to climate change in Australia and Great Britain: Interim report.* Gold Coast, Australia: National Climate Change Adaptation Research Facility.

Revell, K. M. A., & Stanton, N. A. (2012). Models of models: Filtering and bias rings in depiction of knowledge structures and their implications for design. *Ergonomics, 55*(9), 1073–1092.

Reyes, M. L., & Lee, J. D. (2008). Effects of cognitive load presence and duration on driver eye movements and event detection performance. *Transportation Research Part F: Traffic Psychology and Behaviour, 11*(6), 391–402.

Russo, J. E., Johnson, E. J., & Stephens, D. L. (1989). The validity of verbal protocols. *Memory & Cognition, 17*(6), 759–769.

Ryle, G. (1949). *The concept of mind.* London: Hutchinson.

Sanderson, P. M. (2006). The multimodal world of medical monitoring displays. *Applied Ergonomics, 37*(4), 501–512.

Sanderson, P. M., Anderson, J., & Watson, M. O. (2000). Extending ecological interface design to auditory displays. In *Proceedings of the 2000 Annual Conference of the Computer-Human Interaction Special Interest Group (CHISIG) of the Ergonomics Society of Australia (OzCHI2000)*, December 4–8 (pp. 259–266). Sydney: CSIRO.

Sanderson, P. M., Liu, D., & Jenkins, S. A. (2009). Auditory displays in anesthesiology. *Current Opinion in Anaesthesiology, 22*(6), 788–795.

Sanderson, P. M., Naikar, N., Lintern, G., & Goss, S. (1999). Use of cognitive work analysis across the system life-cycle: From requirements to decommissioning. In *Proceedings of the 43rd Annual Meeting of the Human Factors and Ergonomics Society*, Westin Galleria, Houston, Texas, September 27–October 1 (pp. 318–322). Santa Monica, CA: Human Factors and Ergonomics Society.

Sanderson, P. M., & Watson, M. O. (2005). From information content to auditory display with ecological interface design: Prospects and challenges. *Proceedings of the Human Factors and Ergonomics Society Annual Meeting, 49*(3), 259–263.

Sawaragi, T., Shiose, T., & Akashi, G. (2000). Foundations for designing an ecological interface for mobile robot teleoperation. *Robotics and Autonomous Systems, 31*(3), 193–207.

Schmitt, N. (1996). Uses and abuses of coefficient alpha. *Psychological Assessment, 8*(4), 350–353.

Schneiderman, B. (1983). Direct manipulation: A step beyond programming languages. *IEEE Computer, 16*, 57–69.

Schwarzkopf, A. B., & Leipnik, R. B. (1977). Control of highway vehicles for minimum fuel consumption over varying terrain. *Transportation Research, 11*(4), 279–286.

Seaborn, K., & Antle, A. N. (2011). Gauntlet guide: Designing a wearable vibrotactile feedforward display for novice gameplay. *In GRAND '11, May 12–14, 2011.* Vancouver, BC.

Segall, N., Kaber, D. B., Taekman, J. M., & Wright, M. C. (2013). A cognitive modeling approach to decision support tool design for anesthesia provider crisis management. *International Journal of Human-Computer Interaction, 29*(2), 55–66.

Seppelt, B. D., & Lee, J. D. (2007). Making adaptive cruise control (ACC) limits visible. *International Journal of Human-Computer Studies, 65*(3), 192–205.

Seppelt, B. D., Lees, M. N., Lee, J. D., Stoner, H., Brown, M., & Hoffman, J. D. (2005). *Integrating ecological interface design into the engineering design process for vehicles. Nissan Final Report, Year 3.* Nissan, Iowa City.

Shahab, Q., & Terken, J. (2012). Speed vs. acceleration advice for advisory cooperative driving. In N. A. Stanton (Ed.), *Advances in human aspects of road and rail transportation* (pp. 67–86). CRC Press, Boca Raton, Florida.

Sharp, T. D., & Helmicki, A. J. (1998). The application of the ecological interface design approach to neonatal intensive care medicine. In *Proceedings of the Human Factors and Ergonomics Society 42nd Annual Meeting*, Hyatt Regency Chicago, Chicago, Illinois, October 5–9 (pp. 350–354).

Shayler, P. J., Chick, J. P., & Eade., D. (1999). *A method of predicting brake specific fuel consumption maps. No. 1999-01-0556.* SAE Technical Paper.

Sivak, M., & Schoettle, B. (2011). *Eco-driving: Strategic, tactical, and operational decisions of the driver that improve vehicle fuel economy.* Ann Arbor, University of Michigan Transportation Research Institute, MI.

Skoglund, T., & Karlsson, I. C. M. (2012). Appreciated–but with a fading grace of novelty! Traveller's assessment of, usage of and behavioural change given access to a co-modal travel planner. *Procedia – Social and Behavioral Sciences*, *48*, 932–940.

Society of Motor Manufacturers and Traders. (2016). *Registration – EVs and AFVs*. Retrieved January 20, 2016, from http://www.smmt.co.uk/category/news-registration-evs-afvs/

Spence, C., & Ho, C. (2009). Crossmodal information processing in driving. In C. Castro (Ed.), *Human factors of visual and cognitive performance in driving* (pp. 187–200). Boca Raton, FL: CRC Press.

Stanton, N. A., & Baber, C. (1998). Designing for consumers: Editorial. *Applied Ergonomics*, *29*(1), 1–3.

Stanton, N. A., Dunoyer, A., & Leatherland, A. (2011). Detection of new in-path targets by drivers using stop & go adaptive cruise control. *Applied Ergonomics*, *42*(4), 592–601.

Stanton, N. A., & McIlroy, R. C. (2012). Designing mission communication planning: The role of rich pictures and cognitive work analysis. *Theoretical Issues in Ergonomics Science*, *13*(2), 146–168. doi:10.1080/1463922X.2010.497197

Stanton, N. A., McIlroy, R. C., Harvey, C., Blainey, S., Hickford, A., Preston, J. M., & Ryan, B. (2012). Following the cognitive work analysis train of thought: Exploring the constraints of modal shift to rail transport. *Ergonomics*, September, pp. 37–41.

Stanton, N. A., & Stammers, R. B. (2008). Bartlett and the future of ergonomics. *Ergonomics*, *51*(1), 1–13.

Stanton, N. A., & Young, M. S. (2000). A proposed psychological model of driving automation. *Theoretical Issues in Ergonomics Science*, *1*, 315–331.

Stanton, N. A., & Young, M. S. (2005). Driver behaviour with adaptive cruise control. *Ergonomics*, *48*(10), 1294–313.

Stanton, N. A., Young, M. S., Walker, G. H., Turner, H., & Randle, S. (2001). Automating the driver's control tasks. *International Journal of Cognitive Ergonomics*, *5*, 221–236.

Staubach, M., Kassner, A., Fricke, N., Schießl, C., Brockman, M., & Kuck, D. (2012). Driver reactions on ecological driver feedback via different HMI modalities. In *19th ITS World Congress*, 22–26 October, Vienna, Austria.

Staubach, M., Schebitz, N., Fricke, N., Schießl, C., Brockmann, M., & Kuck, D. (2014a). Information modalities and timing of ecological driving support advices. *IET Intelligent Transport Systems*, *8*, 534–542.

Staubach, M., Schebitz, N., Köster, F., & Kuck, D. (2014b). Evaluation of an eco-driving support system. *Transportation Research Part F: Psychology and Behaviour*, *27*, 11–21.

Staubach, M., Schebitz, N., Krehle, T., Oeltze, K., & Kuck, D. (2013). User acceptance of an eco-driving support system. In *9th ITS European Congress*. 4–7 June, Dublin, Ireland.

St-Cyr, O., Jamieson, G. A., & Vicente, K. J. (2013). Ecological interface design and sensor noise. *International Journal of Human-Computer Studies*, *71*(11), 1056–1068.

Steele, M., & Gillespie, R. B. (2001). Shared control between human and machine: Using a haptic steering wheel to aid in land vehicle guidance. In *Proceedings of the Human Factors and Ergonomics Society 45th Annual Meeting*, Minneapolis Hilton and Towers/Minneapolis Convention Center, Minneapolis, Minnesota, October 8–12 (pp. 1671–1675).

Stern, N. (2006). *Stern review: The economics of climate change. Summary of conclusions*. The Stationary Office, London.

Stillwater, T. (2011). *Comprehending consumption: The behavioural basis and implementation of driver feedback for reducing vehicle energy use*. Davis, CA: University of California.

Stillwater, T., & Kurani, K. S. (2011). Field test of energy feedback information: Driver responses and behavioural theory. *Transportation Research Record: Journal of the Transportation Research Board*, *2252*, 7–15.

Stoner, H. A., Wiese, E. E., & Lee, J. D. (2003). Applying ecological interface design to the driving domain: The results of an abstraction hierarchy analysis. In *Proceedings of the Human Factors and Ergonomics Society 47th Annual Meeting*, Adam's Mark Hotel, Denver, Colorado, October 13–17 (Vol. 47, pp. 444–448).

Streiner, D. L. (2003). Starting at the beginning: An introduction to coefficient alpha and internal consistency. *Journal of Personality Assessment, 80*(1), 99–103.

Svenson, O. (1981). Are we all less risky and more skillful than our fellow drivers? *Acta Psychologica, 47*, 143–148.

Swann, W. B., Griffin, J. J., & Predmore, S. C. (1987). The cognitive-affective crossfire: When self-consistency confronts self-enhancement. *Journal of Personality and Social Psychology, 52*(5), 881–889.

Talcott, C. R., Bennett, K. B., Martinez, S. G., & Stansifer, C. (2007). Perception-action icons: An interface design strategy for intermediate domains. *Human Factors, 49*(1), 120–135.

Tango, F., & Montanari, R. (2006). Shaping the drivers' interaction: How the new vehicle systems match the technological requirements and the human needs. *Cognition, Technology & Work, 8*(3), 215–226.

Thatcher, A. (2012). Green ergonomics: Definition and scope. *Ergonomics, 56*(3), 389–398.

The AA. (2015). *Eco-driving advice.* Retrieved January 6, 2016, from http://www.theaa.com/motoring_advice/fuels-and-environment/drive-smart.html

The Shift Project. (2015). *Breakdown of electricity generation by energy source.* Retrieved December 12, 2015, from http://www.tsp-data-portal.org/Breakdown-of-Electricity-Generation-by-Energy-Source#tspQvChart

Thijssen, R., Hofman, T., & Ham, J. (2014). Ecodriving acceptance: An experimental study on anticipation behavior of truck drivers. *Transportation Research Part F: Traffic Psychology and Behaviour, 22*, 249–260.

Thomas, M., Goode, N., Grant, E., Taylor, N. Z., & Salmon, P. M. (2015). Can we talk about speed? The effect of verbal protocols on driver speed and perceived workload. *Procedia Manufacturing, 3*, 2629–2634.

Thull, B., & Rau, G. (1997). Design of an integrated display for hemodynamic monitoring. In *Helmholtz Institute for Biomedical Engineering Research Report 1995/96* (pp. 164–170).

Tijerina, L. (1995). Key human factors research needs in intelligent vehicle-highway system crash avoidance. *Transportation Research Record, 1485*, 1–9.

Torenvliet, G. L., Jamieson, G. A., & Vicente, K. J. (1998). Making the most of ecological interface design: The role of cognitive style. In *Proceedings of the Fourth Annual Symposium on Human Interaction With Complex Systems*. March 22–25, IEEE, Dayton, OH, pp. 214–225.

Travelfootprint.org. (2013). *Reduce your travel footprint – Eco-driving.* Retrieved April 12, 2013, from http://www.travelfootprint.org/ecodriving

Trommer, S., & Höltl, A. (2012). Perceived usefulness of eco-driving assistance systems in Europe. *IET Intelligent Transport Systems, 6*(2), 145–152.

Turrentine, T., Garas, D., Lentz, A., & Woodjack, J. (2011). *The UC Davis MINI E consumer study.* Eds., Fabio Paternò, Maria Francesca Costabile. Institute of Transportation Studies Davis, California.

Underwood, G., Chapman, P., Bowden, K., & Crundall, D. (2002). Visual search while driving: Skill and awareness during inspection of the scene. *Transportation Research Part F: Traffic Psychology and Behaviour, 5*(2), 87–97.

United Nations. (2013). *United Nations Environment Programme: Ecodriving.* Retrieved May 15, 2013, from http://www.unep.org/transport/Programmes/Ecodriving/

Upton, C., & Doherty, G. (2005). Adapting the ADS for high volume manufacturing. In *Human-Computer Interaction – INTERACT 2005* (Vol. M, pp. 1038–1041). Berlin: Springer.

Upton, C., & Doherty, G. (2006a). Designing usable charts for complex work settings. In *Tenth International Conference on Information Visualisation 2006, IV 2006*, 5–7 July (pp. 447–452). IEEE, London, UK.

Upton, C., & Doherty, G. (2006b). Visual representation of complex information structures in high volume manufacturing. In T. Clemmensen, P. Campos, R. Omgreen, A. L. Petjersen, & W. Wong (Eds.), *IFIP International Federation for Information Processing, Volume 221, Human Work Interaction Design: Designing for Human Work* (Vol. M, pp. 27–45). Boston, MA: Springer.

Upton, C., & Doherty, G. (2007). Integrating the visualisation reference model with ecological interface design. In *Proceedings of the ECCE 2007 Conference*, August 28–31, London, UK (pp. 175–178).

Upton, C., & Doherty, G. (2008). Extending ecological interface design principles: A manufacturing case study. *International Journal of Human-Computer Studies, 66*(4), 271–286.

Van Dam, S. B. J., Abeloos, L. M., Mulder, M., & van Paassen, M. M. (2004). Functional presentation of travel opportunities in flexible use airspace: An EID of an airborne conflict support tool. In *IEEE International Conference on Systems, Man, and Cybernetics*, 10–13 Oct, The Hague, The Netherlands (pp. 802–808).

Van Dam, S. B. J., Mulder, M., & van Paassen, M. M. (2005). Ecological interface design of airborne conflict support in flexible use airspace. In *AIAA Guidance, Navigation and Control Conference and Exhibit*. August 15–18, San Francisco, CA.

Van Dam, S. B. J., Mulder, M., & van Paassen, M. M. (2007). Airborne self-separation display with turn dynamics and intruder intent-information. In *IEEE International Conference on Systems, Man and Cybernetics*, October 7–10, Montreal, Quebec, Canada (pp. 1445–1451).

Van Dam, S. B. J., Mulder, M., & van Paassen, M. M. (2008). Ecological interface design of a tactical airborne separation assistance tool. *IEEE Transactions on Systems, Man, and Cybernetics – Part A: Systems and Humans, 38*, 1221–1233.

Van Dam, S. B. J., Steens, C. L. A., Mulder, M., & van Paassen, M. M. (2008). Evaluation of two pilot self-separation displays using conflict situation measurements. In *2008 IEEE International Conference on Systems, Man and Cybernetics*, October 12–15 (pp. 3558–3563). IEEE, Singapore.

Van den Haak, M., De Jong, M., & Jan Schellens, P. (2003). Retrospective vs. concurrent think-aloud protocols: Testing the usability of an online library catalogue. *Behaviour & Information Technology, 22*(5), 339–351.

Van Der Laan, J. D., Heino, A., & De Waard, D. (1997). A simple procedure for the assessment of acceptance of advanced transport telematics. *Transportation Research Part C: Emerging Technologies, 5*(1), 1–10.

Van der Voort, M., Dougherty, M. S., & van Maarseveen, M. (2001). A prototype fuel-efficiency support tool. *Transportation Research Part C: Emerging Technologies, 9*, 279–296.

Van Erp, J. B., & Van Veen, H. A. H. C. (2001). Vibro-tactile information presentation in automobiles. In *Proceedings of Eurohaptics* (Vol. 2001, pp. 99–104). Paris, France: Eurohaptics Society.

Van Erp, J. B. F., & Van Veen, H. A. H. C. (2004). Vibrotactile in-vehicle navigation system. *Transportation Research Part F: Traffic Psychology and Behaviour, 7*(4–5), 247–256.

Van Gog, T., Kester, L., Nievelstein, F., Giesbers, B., & Paas, F. (2009). Uncovering cognitive processes: Different techniques that can contribute to cognitive load research and instruction. *Computers in Human Behaviour, 25*, 325–331.

Van Marwijk, B. J. A., Borst, C., Mulder, M., & van Paasen, M. M. (2011). The international journal of supporting 4D trajectory revisions on the flight deck: Design of a human – Machine interface. *The International Journal of Aviation Psychology, 21*, 35–61.

Van Paassen, M. M., Borst, C., Klomp, R., Mulder, M., van Leeuwen, P., & Mooij, M. (2013). Designing for shared cognition in air traffic management. *Journal of Aerospace Operations, 2,* 39–51.

Várhelyi, A., Hjälmdahl, M., Hydén, C., & Draskóczy, M. (2004). Effects of an active accelerator pedal on driver behaviour and traffic safety after long-term use in urban areas. *Accident Analysis and Prevention, 36,* 729–737.

Varnhagen, R., & Korthaus, C. (2010). *Reduction of fuel consumption with intelligent use of navigation data.* SAE Technical Paper 2010-01-2004.

Vicente, K. J. (1996). Improving dynamic decision making in complex systems through ecological interface design: A research overview. *System Dynamics Review, 12*(4), 251–279.

Vicente, K. J. (1997). Operator adaption in process control: A three-year research program. *Control Engineering Practice, 5*(3), 407–416.

Vicente, K. J. (1999). *Cognitive work analysis: Towards safe, productive and healthy computer-based work.* Mahwah NJ: Lawrence Erlbaum Associates.

Vicente, K. J. (2002). Ecological interface design: Progress and challenges. *Human Factors, 44,* 62–78.

Vicente, K. J., Christoffersen, K., & Pereklita, A. (1995). Supporting operator problem solving through ecological interface design. *IEEE Transactions on Systems, Man, and Cybernetics, 25*(4), 529–545.

Vicente, K. J., Moray, N., Lee, J. D., Rasmussen, J., Jones, B. G., Brock, R., & Djemil, R. (1996). Evaluation of a rankine cycle display for nuclear power plant monitoring and diagnosis. *Human Factors, 38,* 506–521.

Vicente, K. J., & Rasmussen, J. (1990). The ecology of human-machine systems II: Mediating 'direct perception' in complex work domains. *Ecological Psychology, 2*(3), 207–249.

Vicente, K. J., & Rasmussen, J. (1992). Ecological interface design: Theoretical foundations. *IEEE Transactions on Systems, Man, and Cybernetics, 22,* 589–600.

Vignola, R., Klinsky, S., Tam, J., & McDaniels, T. (2012). Public perception, knowledge and policy support for mitigation and adaption to climate change in Costa Rica: Comparisons with North American and European studies. *Mitigation and Adaptation Strategies for Global Change, 18*(3), 303–323.

Wada, T., Yoshimura, K., Doi, S., Youhata, H., & Tomiyama, K. (2011). Proposal of an eco-driving assist system adaptive to driver's skill. In *14th International IEEE Conference on Intelligent Transportation Systems,* 05–07 Oct (pp. 1880–1885). Washington, DC.

Walker, G. H., Stanton, N. A., & Chowdhury, I. (2013). Self explaining roads and situation awareness. *Safety Science, 56,* 18–28.

Walker, G. H., Stanton, N. A., & Salmon, P. M. (2011). Cognitive compatibility of motorcyclists and drivers. *Lecture Notes in Computer Science, 6781*(3), 214–222.

Walker, G. H., Stanton, N. A., & Young, M. S. (2001a). An on-road investigation of feedback and cognitive processing in naturalistic driving. *International Journal of Cognitive Ergonomics, 5,* 421–444.

Walker, G. H., Stanton, N. A., & Young, M. S. (2001b). Hierarchical task analysis of driving: A new research tool. In M. A. Hanson (Ed.), *Contemporary ergonomics – Proceedings of the Annual Conference of the Ergonomics Society, April 2001* (pp. 435–440). London: Taylor and Francis.

Warriner, G. K., McDougall, G. H. G., & Claxton, J. D. (1984). Any data or none at all? Living with inaccuracies in self-reports of residential energy consumption. *Environment and Behavior, 16,* 503–526.

Waters, M. H. L., & Laker, I. B. (1980). *Research on fuel conservation for cars. Report No. 921.* Transport Research Laboratory, Crowthorne, England.

Watson, M. O., Russell, W. J., & Sanderson, P. M. (2000). Ecological interface design for anaesthesia monitoring. *Australian Journal of Information Systems, 7*(2), 109–114.

Watson, M. O., & Sanderson, P. M. (2007). Designing for attention with sound: Challenges and extensions to ecological interface design. *Human Factors: The Journal of the Human Factors and Ergonomics Society, 49*(2), 331–346.

Watson, M. O., Sanderson, P. M., & Anderson, J. (2000). Designing auditory displays for team environments. In *Proceedings of the Fifth Australian Aviation Psychology Symposium*, Manly, Australia. November 20–24.

Wellings, T., Binnersley, J., Robertson, D., & Khan, T. (2011). *Human machine interfaces in low carbon vehicles: Market trends and user issues*.

Wickens, C. D. (2008). Multiple resources and mental workload. *Human Factors, 50*(3), 449–455.

Wickens, C. D., & Carswell, C. M. (1995). The proximity compatibility principle: Its psychological foundation and relevance to display design. *Human Factors: The Journal of the Human Factors and Ergonomics Society, 37*, 473–494.

Wickens, C. D., & Carswell, C. M. (1997). Information processing. In G. Salvendy (Ed.), *Handbook of human factors and ergonomics* (2nd ed., pp. 89–129). New York: Wiley.

Wickens, C. D., Goh, J., Helleberg, J., Horrey, W. J., & Talleur, D. A. (2003). Attentional models of multitask pilot performance using advanced display technology. *Human Factors: The Journal of the Human Factors and Ergonomics Society, 45*(3), 360–380.

Wilson, K. M. (1995). Mass media as sources of global warming knowledge. *Mass Communications Review, 22*, 75–89.

World Commission on Environment and Development. (1987). *Our common future*. Oxford: Oxford University Press.

Wright, H., Mathers, C., & Walton, J. P. R. B. (2013). Using visualization for visualization: An ecological interface design approach to inputting data. *Computers & Graphics, 37*(3), 202–213.

Wu, C., Zhao, G., & Ou, B. (2011). A fuel economy optimization system with applications in vehicles with human drivers and autonomous vehicles. *Transportation Research Part D: Transport and Environment, 16*(7), 515–524.

Xu, W., Dainoff, M. J., & Mark, L. S. (1999). Facilitate complex search tasks in hypertext by externalizing functional properties of a work domain. *International Journal of Human-Computer Interaction, 11*, 201–229.

Young, K. L., Lenné, M. G., Beanland, V., Salmon, P. M., & Stanton, N. A. (2015). Where do novice and experienced drivers direct their attention on approach to urban rail level crossings? *Accident; Analysis and Prevention, 77C*, 1–11.

Young, K. L., Salmon, P. M., & Cornelissen, M. (2013). Distraction-induced driving error: An on-road examination of the errors made by distracted and undistracted drivers. *Accident Analysis and Prevention, 58*, 218–225.

Young, M. S., & Birrell, S. A. (2012). Ecological IVIS design: Using EID to develop a novel in-vehicle information system. *Theoretical Issues in Ergonomics Science, 13*, 225–239.

Young, M. S., Birrell, S. A., & Stanton, N. A. (2011). Safe driving in a green world: A review of driver performance benchmarks and technologies to support 'smart' driving. *Applied Ergonomics, 42*(4), 533–539.

Yu, H., Wang, B., Zhang, Y.-J., Wang, S., & Wei, Y.-M. (2013). Public perception of climate change in China: Results from the questionnaire survey. *Natural Hazards, 69*(1), 459–472.

Zachrisson, J., & Boks, C. (2010). When to apply different design for sustainable behaviour strategies? In *ERSCP-EMSU Conference on Knowledge Collaboration & Learning for Sustainable Innovation*, October 25–29. Delft, The Netherlands.

Appendix A: NASA Task Load Index – Raw TLX

Mental demand   How mentally demanding was the task?

Very low   Very high

Physical demand   How physically demanding was the task?

Very low   Very high

Temporal demand   How hurried or rushed was the pace of the task?

Very low   Very high

Performance   How successful were you in accomplishing what you were asked to do?

Perfect   Failure

Effort   How hard did you have to work to accomplish your level of performance?

Very low   Very high

Frustration   How insecure, discouraged, irritated, stressed and annoyed were you?

Very low   Very high

Appendix B: Van Der Laan Acceptance Scale

My judgements of the information system are... (please tick a box on every line)

Useful						Useless
Pleasant						Unpleasant
Bad						Good
Nice						Annoying
Effective						Superfluous
Irritating						Likeable
Assisting						Worthless
Undesirable						Desirable
Raising alertness						Sleep inducing

Appendix C: Eco-driving Survey – Driving and the Environment

SECTION 1. INFORMATION AND CONSENT

QUESTION 1.1

We are interested in fuel use in the private road vehicle (i.e. the car), and the purpose of this survey is to gather information regarding people's knowledge of and attitudes towards driving and the environment. This research is funded partly by the Engineering and Physical Sciences Research Council and partly by Jaguar Land Rover.

You have been approached to take part in this study as someone who is 18 or over and has at least 1 year's experience in driving. Your participation in this research study is entirely voluntary. You may choose not to participate. If you decide to participate in this research survey, you may withdraw at any time. If you decide not to participate in this study, or if you withdraw from participating at any time, you will not be penalised.

Should you choose to take part, you will be asked to fill out a questionnaire that should take approximately 15 minutes. The questions are about your general attitudes towards the environment, about your driving habits and about your knowledge of and attitudes towards fuel-efficient driving. Though there are no significant benefits to your participating in this research, there are also no risks beyond that experienced in normal, day-to-day life.

This research has been reviewed according to the University of Southampton ethics procedures for research involving human participants.

Your continued participation in this research will be taken as evidence of your giving informed consent to participate in this study and for your data to be used for the purposes of research, and that you understand that published results of this research project will maintain your confidentially. Your participation is voluntary and you may withdraw your participation at any time.

SECTION 2. ABOUT YOU

QUESTION 2.1

What is your gender?
- O Male
- O Female

QUESTION 2.2

What is your age?
- 18–24
- 25–34
- 35–44
- 45–54
- 55–64
- 65 or over

QUESTION 2.3

In which country do you live?

QUESTION 2.4

What is the highest level of education you have completed?
- O level, CSE, GCSE or equivalent
- A level or equivalent
- Undergraduate degree (e.g. BSc, BA)
- Postgraduate degree (e.g. MSc, MRes, PhD)
- None of the above

QUESTION 2.5

In which year did you get your driving licence? (Please enter 4-digit year, e.g. 1998)

QUESTION 2.6

What type of driving licence(s) do you have? (Tick all that apply)
- Moped (up to 50cc engine)
- Motorcycle
- Car
- Medium-sized vehicle (3.5 to 7.5 tonnes with trailer up to 12 tonnes total)
- Large vehicle (over 3.5 tonnes)
- Minibus (up to 16 passengers, 8 meters length)
- Bus (any bus with over 8 passenger seats)

QUESTION 2.7

Do you own or have regular access to a vehicle (e.g. one in the household or at the workplace?)
- Yes
- No

QUESTION **2.8**

Are you a professional driver, that is, do you drive for your work?
   ○ Yes as the main part of my job
   ○ Yes but not as the main part of my job
   ○ No

QUESTION **2.9**

Are you a fleet driver?
   ○ Yes
   ○ No

QUESTION **2.10**

In a typical week, on how many days do you drive?
   ○ I don't drive every week
   ○ 1 or 2 days a week
   ○ 3 or 4 days a week
   ○ 5 or 6 days a week
   ○ Everyday

QUESTION **2.11**

In a typical week (i.e. Monday to Friday) how long are your drives?
   ○ Less than 1 mile (less than 0.7 km)
   ○ 1 to 3 miles (0.7 to 5 km)
   ○ 3 to 10 miles (5 to 16 km)
   ○ 10 to 50 miles (16 to 80 km)
   ○ More than 50 miles (more than 80 km)
   ○ I typically don't drive in the week

QUESTION **2.12**

In a typical weekend how long are your drives?
   ○ Less than 1 mile (less than 0.7 km)
   ○ 1 to 3 miles (0.7 to 5 km)
   ○ 3 to 10 miles (5 to 16 km)
   ○ 10 to 50 miles (16 to 80 km)
   ○ More than 50 miles (more than 80 km)
   ○ I typically don't drive on the weekend

QUESTION **2.13**

Approximately how many miles have you driven in the last year?
   ○ Less than 1000 miles (less than 1600 km)
   ○ 1000 to 5000 miles (1600 to 8000 km)

○ 5000 to 10,000 miles (8000 to 16,000 km)
○ 10,000 to 15,000 miles (16,000 to 24,000 km)
○ More than 15,000 miles (more than 24,000 km)

Question 2.14

On which types of road do you most often drive? (please select one or two)
○ Motorways and dual carriageways
○ Rural (e.g. country roads)
○ Urban (i.e. around town)

Question 2.15

Have you ever attended an advance or additional driver-training course (beyond initial pre-test driver training)?
○ Yes
○ No

Question 2.16

Do you belong to a motoring organisation (e.g. IAM or RoSPA)?
○ Yes
○ No

Question 2.17

If so, which organisation?

Question 2.18a

What type of vehicle do you drive for personal use (i.e. not work – though they can be the same vehicle)?
○ Moped or motorcycle
○ Car or van
○ Other

Question 2.18b

If 'other', please specify the type (e.g. bus, lorry/truck)

Question 2.19

What is the make and model of this vehicle?
Make [   ]
Model [   ]

QUESTION **2.20**

What type of fuel does this vehicle use?
- ○ Petrol
- ○ Diesel
- ○ Electricity and other fuel (i.e. a hybrid vehicle)
- ○ Electricity only (i.e. battery electric vehicle)
- ○ LPG
- ○ CNG
- ○ Biofuel
- ○ Other
- ○ Don't know

QUESTION **2.21**

What is the engine size of this vehicle?
- ○ Less than 1 L (1000cc)
- ○ 1 to 1.2 L
- ○ 1.21 to 14 L
- ○ 1.41 to 1.6 L
- ○ 1.61 to 1.8 L
- ○ 1.81 to 2 L
- ○ 2.01 to 2.2 L
- ○ 2.21 to 2.4 L
- ○ 2.41 to 2.6 L
- ○ 2.61 to 2.8 L
- ○ 2.81 to 3 L
- ○ 3.01 to 3.2 L
- ○ 3.21 to 3.4 L
- ○ 3.41 to 3.6 L
- ○ 3.61 to 3.8 L
- ○ 3.81 to 4 L
- ○ 4.01 to 4.2 L
- ○ 4.21 to 4.4 L
- ○ 4.41 to 4.6 L
- ○ 4.61 to 4.8 L
- ○ 4.81 to 5 L
- ○ More than 5 L
- ○ Unknown

QUESTION **2.22**

In what year was this vehicle first registered? (Please enter 4-digit year, e.g. 2008) If unknown please leave blank

QUESTION 2.23

With this vehicle, about how many miles per gallon do you typically get? (do not worry about being exact) If completely unknown, please leave blank

QUESTION 2.24

Different vehicles have different average fuel efficiencies – could you provide an estimate in miles per gallon, of the fuel efficiencies of the following vehicles (assuming all run on petrol)?

Up to 125cc Motorcycle	[ ]
126cc plus Motorcycle	[ ]
Small car (under 1.7 L engine)	[ ]
Medium car (1.7 to 2.7 L engine)	[ ]
Large car (over 2.7 L engine)	[ ]
People mover or van	[ ]
SUV or 4-wheel drive over 2.7 L	[ ]
Truck bus or campervan	[ ]

QUESTION 2.25

The way in which a car is driven affects the amount of fuel consumed per mile– about how much difference do you think this 'driving behaviour' can have for the average person?

- O 0–5%
- O 5–10%
- O 10–15%
- O 15–20%
- O 20–25%
- O 25–30%
- O 30–35%
- O More than 35%

QUESTION 2.26

What kind of effect do you think it could have for your fuel use?

- O 0–5%
- O 5–10%
- O 10– 15%
- O 15–20%
- O 20–25%
- O 25–30%
- O 30–35%
- O More than 35%

QUESTION 2.27

Have you heard about the practice of 'eco-driving'?
- ○ Yes I know what it means and I am confident that I know how to do it
- ○ Yes and I know what it means and I have an idea of how to do it
- ○ Yes I know what it means but don't know how to do it
- ○ Yes but I have only heard of it and am not sure what it means
- ○ No

QUESTION 2.28

What do you think of 'eco-driving'? (Please tick all that apply)
- ○ A good idea
- ○ Good for the environment
- ○ Helps drivers save money
- ○ The UK/the world doesn't need it
- ○ I'm too busy to worry about it
- ○ Reduces driving enjoyment too much
- ○ Time pressure is more important than fuel use
- ○ It is unsafe
- ○ Don't know/haven't heard of it

QUESTION 2.29

Could you give a tip for reducing fuel consumption while driving? (Skip to question 37 if not)

QUESTION 2.30

How often do you follow this advice?
- ○ Always or almost always
- ○ Usually
- ○ Sometimes
- ○ Rarely
- ○ Never or almost never

QUESTION 2.31

Could you give another tip for reducing fuel consumption while driving? (Skip to question 37 if not)

QUESTION 2.32

How often do you follow this advice?
- ○ Always or almost always
- ○ Usually
- ○ Sometimes

○ Rarely
○ Never or almost never

Question 2.33

Could you give another tip for reducing fuel consumption while driving? (Skip to question 37 if not)

Question 2.34

How often do you follow this advice?
○ Always or almost always
○ Usually
○ Sometimes
○ Rarely
○ Never or almost never

Question 2.35

Could you give another tip for reducing fuel consumption while driving? (Skip to question 37 if not)

Question 2.36

How often do you follow this advice?
○ Always or almost always
○ Usually
○ Sometimes
○ Rarely
○ Never or almost never

Question 2.37

How much would you be prepared to pay for a professional training course to improve your driving style (in terms of fuel efficiency)?
○ Nothing
○ Up to £50
○ £50 to £100
○ £100 to £200
○ £200 to £300
○ £300 or more

Question 2.38

Are you aware that there are in-vehicle devices that aim to help people improve their fuel efficiency while driving (power gauges, 'ecometers', etc.)?
○ No
○ Yes, but my vehicle does not have one

    ○ Yes, and my vehicle has one but I do not use it
    ○ Yes, and my vehicle has one that I have used occasionally
    ○ Yes, and my vehicle has one that I use often

QUESTION 2.39

If you have experience with such a system, do you think it has helped you to save fuel when driving?
    ○ Yes
    ○ No
    ○ Don't know

QUESTION 2.40

If you do not have experience with such a system, do you think it would help you save fuel if you did have one?
    ○ Yes
    ○ No
    ○ Don't know

QUESTION 2.41

How much would you be willing to pay for such a system?
    ○ Nothing
    ○ Up to £50
    ○ £50 to £100
    ○ £100 to £200
    ○ £200 to £300
    ○ £300 or more

SECTION 3. THE ENVIRONMENT

QUESTION 3.1

Could you please tick one box on each line to indicate how much you agree with the following statements.

I am motivated to save money on energy consumption at home	○	○	○	○	○	○	○
We need to find better ways to produce clean and safe energy	○	○	○	○	○	○	○
People at work don't care too much about saving fuel or energy	○	○	○	○	○	○	○
We live in an energy-guzzling society	○	○	○	○	○	○	○
I think that energy issues are	○	○	○	○	○	○	○
My own contribution to saving energy and fuel could be better	○	○	○	○	○	○	○

I believe high energy consumption is bad for the
environment   O O O O O O O

I do not see how, in this country, we can make large
reductions in our fuel and energy use   O O O O O O O

I would travel by public transport more if it were
cheaper than it is presently   O O O O O O O

When driving my own car, I like to keep a check on
the miles per gallon   O O O O O O O

I get from that car   O O O O O O O

When I next buy a car, I will choose one with better
fuel consumption than my current car   O O O O O O O

Energy prices will have to rise quite a lot if we are to
sort out environmental problems   O O O O O O O

People will only change their energy-consuming
habits if they are forced to   O O O O O O O

I am concerned that gas and oil for fuel will run out
in the next 30 years   O O O O O O O

At home, I make sure I get the cheapest energy
possible   O O O O O O O

I try to reduce energy consumption generally at home   O O O O O O O

People at work would not generally try to save energy
unless there was some incentive to do so   O O O O O O O

I switch lights off wherever and whenever I see them
on but not being used   O O O O O O O

It is important to complete a journey as quickly as
possible   O O O O O O O

It bothers me that sometimes in the city centre you
can smell diesel and petrol fumes   O O O O O O O

I would support traffic congestion charging as a
means to reduce traffic jams and pollution   O O O O O O O

It annoys me when people waste energy   O O O O O O O

I would travel on public transport more if it were
more convenient   O O O O O O O

It is a waste of time trying to get people to use cars
less   O O O O O O O

I would only buy a more eco-friendly car if it was no
more expensive to buy than any other car   O O O O O O O

People care more about saving fuel at home than at
work   O O O O O O O

Thank you for taking this questionnaire.   O O O O O O O

Appendix D: Environmental Attitudes Inventory – Short

1. I really like going on trips into the countryside, for example, to forests or fields.

1	2	3	4	5	6	7
Strongly disagree						Strongly agree

2. I find it very boring being out in wilderness areas.

1	2	3	4	5	6	7
Strongly disagree						Strongly agree

3. Being out in nature is a great stress reducer for me.

1	2	3	4	5	6	7
Strongly disagree						Strongly agree

4. I have a sense of well-being in the silence of nature.

1	2	3	4	5	6	7
Strongly disagree						Strongly agree

5. I find it more interesting in a shopping mall than out in the forest looking at trees and birds.

1	2	3	4	5	6	7
Strongly disagree						Strongly agree

6. I think spending time in nature is boring.

1	2	3	4	5	6	7
Strongly disagree						Strongly agree

7. Governments should control the rate at which raw materials are used to ensure that they last as long as possible.

1	2	3	4	5	6	7
Strongly disagree						Strongly agree

8. Controls should be placed on industry to protect the environment from pollution, even if it means things will cost more.

1	2	3	4	5	6	7
Strongly disagree						Strongly agree

9. People in developed societies are going to have to adopt a more conserving lifestyle in the future.

1	2	3	4	5	6	7

Strongly disagree   Strongly agree

10. I don't think people in developed societies are going to have to adopt a more conserving lifestyle in the future.

1	2	3	4	5	6	7

Strongly disagree   Strongly agree

11. Industries should be able to use raw materials rather than recycled ones if this leads to lower prices and costs, even if it means the raw materials will eventually be used up.

1	2	3	4	5	6	7

Strongly disagree   Strongly agree

12. I am opposed to governments controlling and regulating the way raw materials are used in order to try and make them last longer.

1	2	3	4	5	6	7

Strongly disagree   Strongly agree

13. I would like to join and actively participate in an environmentalist group.

1	2	3	4	5	6	7

Strongly disagree   Strongly agree

14. I don't think I would help to raise funds for environmental protection.

1	2	3	4	5	6	7

Strongly disagree   Strongly agree

15. I would NOT get involved in an environmentalist organisation.

1	2	3	4	5	6	7

Strongly disagree   Strongly agree

16. Environmental protection costs a lot of money. I am prepared to help out in a fund-raising effort.

1	2	3	4	5	6	7

Strongly disagree   Strongly agree

17. I would not want to donate money to support an environmentalist cause.

1	2	3	4	5	6	7

Strongly disagree   Strongly agree

18. I would like to support an environmental organisation.

    1   2   3   4   5   6   7

Strongly disagree   Strongly agree

19. One of the most important reasons to keep lakes and rivers clean is that people have a place to enjoy water sports.

    1   2   3   4   5   6   7

Strongly disagree   Strongly agree

20 Nature is important because of what it can contribute to the pleasure and welfare of humans.

    1   2   3   4   5   6   7

Strongly disagree   Strongly agree

21. The thing that concerns me most about deforestation is that there will not be enough lumber for future generations.

    1   2   3   4   5   6   7

Strongly disagree   Strongly agree

22. Conservation is important even if it lowers peoples' standard of living.

    1   2   3   4   5   6   7

Strongly disagree   Strongly agree

23. We need to keep rivers and lakes clean in order to protect the environment, and NOT as places for people to enjoy water sports.

    1   2   3   4   5   6   7

Strongly disagree   Strongly agree

24. We should protect the environment even if it means peoples' welfare will suffer.

    1   2   3   4   5   6   7

Strongly disagree   Strongly agree

25. Science and technology will eventually solve our problems with pollution, overpopulation and diminishing resources.

    1   2   3   4   5   6   7

Strongly disagree   Strongly agree

26. Modern science will NOT be able to solve our environmental problems.

    1   2   3   4   5   6   7

Strongly disagree   Strongly agree

27. We cannot keep counting on science and technology to solve our environmental problems.

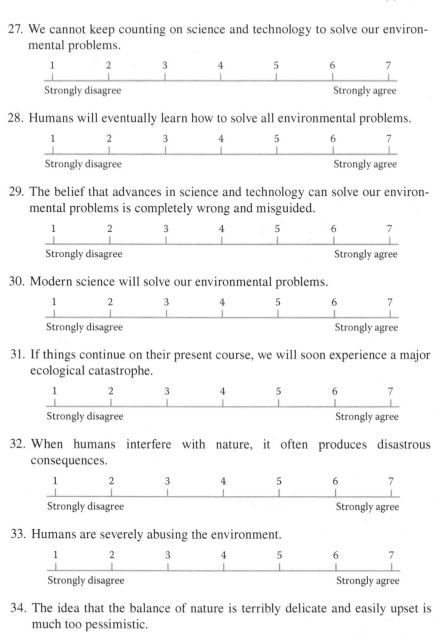

     1   2   3   4   5   6   7
     Strongly disagree   Strongly agree

28. Humans will eventually learn how to solve all environmental problems.

     1   2   3   4   5   6   7
     Strongly disagree   Strongly agree

29. The belief that advances in science and technology can solve our environmental problems is completely wrong and misguided.

     1   2   3   4   5   6   7
     Strongly disagree   Strongly agree

30. Modern science will solve our environmental problems.

     1   2   3   4   5   6   7
     Strongly disagree   Strongly agree

31. If things continue on their present course, we will soon experience a major ecological catastrophe.

     1   2   3   4   5   6   7
     Strongly disagree   Strongly agree

32. When humans interfere with nature, it often produces disastrous consequences.

     1   2   3   4   5   6   7
     Strongly disagree   Strongly agree

33. Humans are severely abusing the environment.

     1   2   3   4   5   6   7
     Strongly disagree   Strongly agree

34. The idea that the balance of nature is terribly delicate and easily upset is much too pessimistic.

     1   2   3   4   5   6   7
     Strongly disagree   Strongly agree

35. I do not believe that the environment has been severely abused by humans.

     1   2   3   4   5   6   7
     Strongly disagree   Strongly agree

36. People who say that the unrelenting exploitation of nature has driven us to the brink of ecological collapse are wrong.

1	2	3	4	5	6	7

Strongly disagree   Strongly agree

37. I'd prefer a garden that is wild and natural to a well groomed and ordered one.

1	2	3	4	5	6	7

Strongly disagree   Strongly agree

38. Human beings should not tamper with nature even when nature is uncomfortable and inconvenient for us.

1	2	3	4	5	6	7

Strongly disagree   Strongly agree

39. Turning new unused land over to cultivation and agricultural development should be stopped.

1	2	3	4	5	6	7

Strongly disagree   Strongly agree

40. I'd much prefer a garden that is well groomed and ordered to a wild and natural one.

1	2	3	4	5	6	7

Strongly disagree   Strongly agree

41. When nature is uncomfortable and inconvenient for humans, we have every right to change and remake it to suit ourselves.

1	2	3	4	5	6	7

Strongly disagree   Strongly agree

42. Grass and weeds growing between pavement stones really look untidy.

1	2	3	4	5	6	7

Strongly disagree   Strongly agree

43. I could not be bothered to save water or other natural resources.

1	2	3	4	5	6	7

Strongly disagree   Strongly agree

44. In my daily life, I'm just not interested in trying to conserve water and/or power.

1	2	3	4	5	6	7

Strongly disagree   Strongly agree

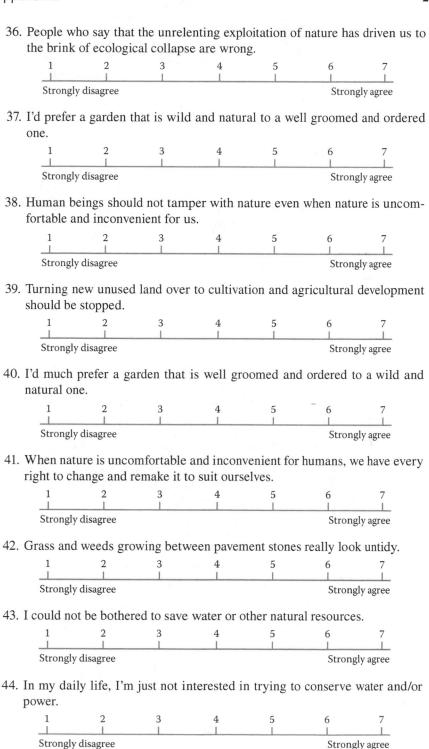

45. I always switch the light off when I don't need it on any more.

| 1 | 2 | 3 | 4 | 5 | 6 | 7 |

Strongly disagree   Strongly agree

46. In my daily life, I try to find ways to conserve water or power.

| 1 | 2 | 3 | 4 | 5 | 6 | 7 |

Strongly disagree   Strongly agree

47. I am NOT the kind of person who makes efforts to conserve natural resources.

| 1 | 2 | 3 | 4 | 5 | 6 | 7 |

Strongly disagree   Strongly agree

48. Whenever possible, I try to save natural resources.

| 1 | 2 | 3 | 4 | 5 | 6 | 7 |

Strongly disagree   Strongly agree

49. Humans were meant to rule over the rest of nature.

| 1 | 2 | 3 | 4 | 5 | 6 | 7 |

Strongly disagree   Strongly agree

50. Human beings were created or evolved to dominate the rest of nature.

| 1 | 2 | 3 | 4 | 5 | 6 | 7 |

Strongly disagree   Strongly agree

51. Plants and animals have as much right as humans to exist.

| 1 | 2 | 3 | 4 | 5 | 6 | 7 |

Strongly disagree   Strongly agree

52. Plants and animals exist primarily to be used by humans.

| 1 | 2 | 3 | 4 | 5 | 6 | 7 |

Strongly disagree   Strongly agree

53. I DO NOT believe humans were created or evolved to dominate the rest of nature.

| 1 | 2 | 3 | 4 | 5 | 6 | 7 |

Strongly disagree   Strongly agree

54. Humans are no more important than any other species.

| 1 | 2 | 3 | 4 | 5 | 6 | 7 |

Strongly disagree   Strongly agree

55. Protecting peoples' jobs is more important than protecting the environment.

1	2	3	4	5	6	7

Strongly disagree   Strongly agree

56. Humans do NOT have the right to damage the environment just to get greater economic growth.

1	2	3	4	5	6	7

Strongly disagree   Strongly agree

57. Protecting the environment is more important than protecting economic growth.

1	2	3	4	5	6	7

Strongly disagree   Strongly agree

58. Protecting the environment is more important than protecting peoples' jobs.

1	2	3	4	5	6	7

Strongly disagree   Strongly agree

59. The question of the environment is secondary to economic growth.

1	2	3	4	5	6	7

Strongly disagree   Strongly agree

60. The benefits of modern consumer products are more important than the pollution that results from their production and use.

1	2	3	4	5	6	7

Strongly disagree   Strongly agree

61. The idea that nature is valuable for its own sake is naïve and wrong.

1	2	3	4	5	6	7

Strongly disagree   Strongly agree

62. Nature is valuable for its own sake.

1	2	3	4	5	6	7

Strongly disagree   Strongly agree

63. I do not believe protecting the environment is an important issue.

1	2	3	4	5	6	7

Strongly disagree   Strongly agree

64. Despite our special abilities, humans are still subject to the laws of nature.

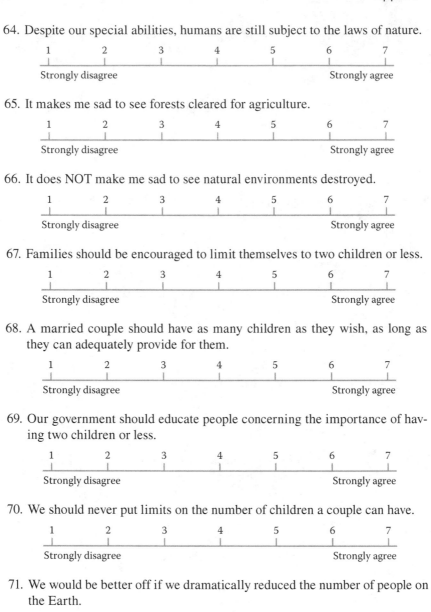

    1   2   3   4   5   6   7

Strongly disagree   Strongly agree

65. It makes me sad to see forests cleared for agriculture.

    1   2   3   4   5   6   7

Strongly disagree   Strongly agree

66. It does NOT make me sad to see natural environments destroyed.

    1   2   3   4   5   6   7

Strongly disagree   Strongly agree

67. Families should be encouraged to limit themselves to two children or less.

    1   2   3   4   5   6   7

Strongly disagree   Strongly agree

68. A married couple should have as many children as they wish, as long as they can adequately provide for them.

    1   2   3   4   5   6   7

Strongly disagree   Strongly agree

69. Our government should educate people concerning the importance of having two children or less.

    1   2   3   4   5   6   7

Strongly disagree   Strongly agree

70. We should never put limits on the number of children a couple can have.

    1   2   3   4   5   6   7

Strongly disagree   Strongly agree

71. We would be better off if we dramatically reduced the number of people on the Earth.

    1   2   3   4   5   6   7

Strongly disagree   Strongly agree

72. The government has no right to require married couples to limit the number of children they can have.

    1   2   3   4   5   6   7

Strongly disagree   Strongly agree

Author Index

Subject Index